MW01599193

Human Disease and Prevention

JONES & BARTLETT
LEARNING

World Headquarters
Jones & Bartlett Learning
5 Wall Street
Burlington, MA 01803
978-443-5000
info@jblearning.com
www.jblearning.com

Jones & Bartlett Learning books and products are available through most bookstores and online booksellers. To contact Jones & Bartlett Learning directly, call 800-832-0034, fax 978-443-8000, or visit our website, www.jblearning.com.

Substantial discounts on bulk quantities of Jones & Bartlett Learning publications are available to corporations, professional associations, and other qualified organizations. For details and specific discount information, contact the special sales department at Jones & Bartlett Learning via the above contact information or send an email to specialsales@jblearning.com.

Copyright © 2013 by Jones & Bartlett Learning, LLC, an Ascend Learning Company

All rights reserved. No part of the material protected by this copyright may be reproduced or utilized in any form, electronic or mechanical, including photocopying, recording, or by any information storage and retrieval system, without written permission from the copyright owner.

This publication is designed to provide accurate and authoritative information in regard to the subject matter covered. It is sold with the understanding that the publisher is not engaged in rendering legal, accounting, or other professional service. If legal advice or other expert assistance is required, the service of a competent professional person should be sought.

Production Credits

Chief Executive Officer: Ty Field
President: James Homer
SVP, Editor-in-Chief: Michael Johnson
SVP, Chief Technology Officer: Dean Fossella
SVP, Chief Marketing Officer: Alison M. Pendergast
Publisher, Higher Education: Cathleen Sether
Associate Production Editor: Tina Chen
Manufacturing and Inventory Control Supervisor: Amy Bacus
Composition: Abella Publishing Services
Cover Design: Carolyn Downer
Photo Research Supervisor: Anna Genoese
Cover Image: Clockwise, from top: Courtesy of James Gathany/CDC; © dgrilla/Fotolia.com; Courtesy of Janice Haney Carr/Jeff Hageman, M.H.S./CDC; Courtesy of Mass Communication Specialist 3rd Class Jake Berenguer/U.S. Navy; back cover: Courtesy of Mass Communication Specialist 3rd Class Jake Berenguer/U.S. Navy
Table of Contents: Page iii, © sergel telegin/ShutterStock, Inc.; page iv, Courtesy of CDC; page v, Courtesy of Janice Carr/CDC; page vi, © Neale Cousland/ShutterStock, Inc.
Opener Images: Chapters 1, 3, 5, 6, Glossary A © sergel telegin/ShutterStock, Inc.; Chapters 2, 8, 9 © Jaimie Duplass/ShutterStock, Inc.; Chapter 4 Courtesy of Mate 3rd Class Alysha Chavez/U.S. Navy; Chapter 7 Courtesy of Ethleen Lloyd/CDC
Printing and Binding: Courier Companies
Cover Printing: Courier Companies

Some images in this book feature models. These models do not necessarily endorse, represent, or participate in the activities represented in the images.

ISBN: 978-1-4496-4807-7

6048
Printed in the United States of America
15 14 13 12 11 10 9 8 7 6 5 4 3 2 1

Contents

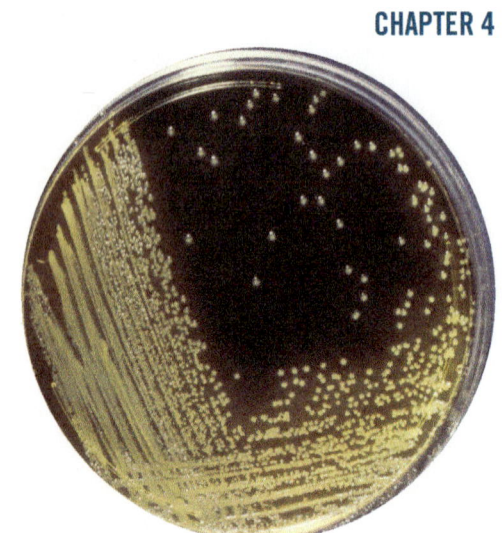

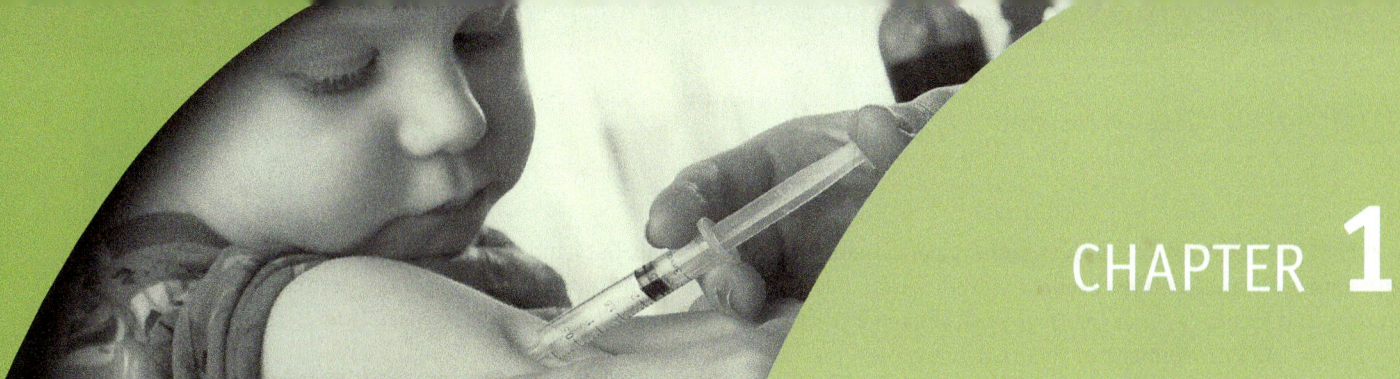

CHAPTER 1

Public Health: The Population Health Approach

LEARNING OBJECTIVES:

By the end of this chapter the student will be able to:

- identify multiple ways that public health affects daily life.
- define eras of public health from ancient times to the early 21st century.
- define the meaning of population health.
- illustrate the uses of health care, traditional public health, and social interventions in population health.
- identify a range of determinants of disease.

I woke up this morning, got out of bed, and went to the bathroom where I used the toilet, washed my hands, brushed and flossed my teeth, drank a glass of water, and took my blood pressure medicine, cholesterol medication, and an aspirin. Then, I did my exercises and took a shower.

On the way to the kitchen, I didn't even notice the smoke detector I passed or the old ashtrays in the closet. I took a low fat yogurt out of the refrigerator and prepared hot cereal in the microwave oven for my breakfast.

Then, I walked out my door into the crisp clean air and got in my car. I put on my seat belt, saw the light go on for the air bag, and safely drove to work. I got to my office where I paid little attention to the new defibrillator at the entrance, the "no smoking" signs, or the absence of asbestos. I arrived safely in my well-ventilated office and got ready to teach Public Health 101.

It wasn't a very eventful morning, but then it's all in a morning's work when it comes to public health.

This rather mundane morning is made possible by a long list of achievements that reflect the often-ignored history of public health.[1] We take for granted the fact that water chlorination, hand washing, and indoor plumbing largely eliminated the transmission of common bacterial disease, which so often killed the young and not-so-young for centuries. Don't overlook the impact of prevention on our teeth and gums. Teeth brushing, flossing, and fluoridation of water have made a dramatic impact on dental health of children and adults.

The more recent advances in the prevention of heart disease have been a major public health achievement. Preventive successes include: the reduction of blood pressure and cholesterol, cigarette cessation efforts, use of low-dose aspirin, an understanding of the role of exercise, and the widespread availability of defibrillators. These can be credited with at least half the dramatic reductions in heart disease that have reduced the death rate from coronary artery disease by approximately 50 percent in the United States and most other developed countries in the last half century.

The refrigerator was one of the most important advances in food safety which illustrates the impact of social change and innovation not necessarily intended to improve health. Food and product safety are public health achievements that require continued attention. It was public pressure for food safety that in large part brought about the creation of the U.S. Food and Drug Administration. The work of this public health agency continues to affect all of our lives from the safety of the foods we eat to the drugs and cosmetics we use.

Radiation safety, like radiation itself, usually goes unnoticed from the regulation of microwave ovens to the reduction of radon in buildings. We rarely notice when disease does not occur.

Highway safety illustrates the wide scope of activities required to protect the public's health. From seat belts, child restraints, and air bags to safer cars, highways, designated driver programs and enforcement of drunk driving laws, public health efforts require collaboration with professionals not usually thought of as having a health focus.

The physical environment too has been made safer by the efforts of public health. Improvement in the quality of the air we breathe both outdoors and indoors has been an ongoing accomplishment of what we will call "population health." Our lives are safer today because of interventions ranging from installation of smoke detectors to removal of asbestos from buildings. However, the challenges continue. Globalization increases the potential for the spread of existing and emerging diseases and raises concerns about the safety of the products we use. Climate change and ongoing environmental deterioration continue to produce new territory for "old" diseases, such as malaria. Overuse of technologies, such as antibiotics, have encouraged the emergence of resistant bacteria.

The 20th century saw an increase in life expectancy of almost 30 years in most developed countries, much of it due to the successes of public health initiatives.[2] We cannot assume that these trends will continue indefinitely. The epidemic of obesity already threatens to slow down or reverse the progress we have been making. The challenges of 21st century public health include protection of health and continued improvement in its quality, not just its quantity.

To understand the role of public health in these achievements and ongoing challenges, let us start at the beginning and ask: what do we mean by public health?

WHAT DO WE MEAN BY PUBLIC HEALTH?

Ask your parents what public health means and they might say "health care for the poor." Well, they are right that public health has always been about providing services for those with special vulnerabilities either directly or through the healthcare system. But that is only one of the ways that public health serves the most needy and vulnerable in our population. Public health efforts often focus on the most vulnerable populations from reducing exposure to lead paint in deteriorating buildings to food supplementation to prevent birth defects and goiters. Addressing the needs of vulnerable populations has always been a cornerstone of public health. As we will see, however, the definition of vulnerable populations continues to change as do the challenges of addressing their needs.

Ask your grandparents what public health means and they might say "washing your hands." Well, they are right too—public health has always been about determining risks to health and providing successful interventions that are applicable to everyone. But hand washing is only the tip of the iceberg. The types of interventions that apply to everyone and benefit everyone span an enormous range: from food and drug safety to controlling air pollution; from measures to prevent the spread of tuberculosis to vaccinating against childhood diseases; from prevention and response to disasters to detection of contaminants in our water.

The concerns of society as a whole are always in the forefront of public health. These concerns keep changing and the methods for addressing them keep expanding. New technologies and global, local, and national interventions are becoming a necessary part of public health. To understand what public health has been and what it is becoming, let us look at some definitions of public health. The following are two definitions of public health—one from the early 20th century and one from more recent years.

Public health is ". . . the science and art of preventing disease, prolonging life and promoting health . . . through organized community effort. . . ."[3]

The substance of public health is the "organized community efforts aimed at the prevention of disease and the promotion of health."[4]

These definitions show how little the concept of public health changed in the 20th century, however the concept of public health in the 21st century is beginning to undergo important changes in a number of ways including:

- The goal of prolonging life is being complemented by an emphasis on the quality of life.
- Protection of health when it already exists is becoming a focus along with promoting health when it is at risk.
- Use of new technologies, such as the Internet, are redefining "community," as well as offering us new ways to communicate.
- The enormous expansion in the options for intervention, as well as the increasing awareness of potential harms and costs of intervention programs, require a new science of "evidence-based" public health.
- Public health and clinical care, as well as public and private partnerships, are coming together in new ways to produce collaborative efforts rarely seen in the 20th century.

Thus, a new 21st century definition of public health is needed. One such definition might read as follows:

The totality of all evidence-based public and private efforts that preserve and promote health and prevent disease, disability, and death.

This broad definition recognizes public health as the umbrella for a range of approaches which need to be viewed as a part of a big picture or population perspective. Specifically, this definition enlarges the traditional scope of public health to include:

- An examination of the full range of environmental, social, and economic determinants of health—not just those traditionally addressed by the public health and clinical health care
- An examination of the full range of interventions to address health issues, including the structure and function of healthcare delivery systems, plus the role of public policies that affect health even when health is not their intended effect

If you are asked by your children what is public health, you might respond: *"It is about the big picture issues that affect our own health and the health of our community every day of our lives. It is about protecting health in the face of disasters; preventing disease from addictions such as cigarettes; controlling infections such as the human immunodeficiency virus (HIV); and developing systems to ensure the safety of the food we eat and the water we drink."*

A variety of terms have been used to describe this big picture perspective that takes into account the full range of factors that affect health and considers their interactions.[5] A variation of this approach has been called the social-ecological model, systems thinking, or the **population health approach**. We will use the latter term. Before exploring what we mean by the population health approach, let us examine how the approaches to public health have changed over time.[a]

HOW HAS THE APPROACH OF PUBLIC HEALTH CHANGED OVER TIME?

Organized community efforts to promote health and prevent disease go back to ancient times.[6,7] The earliest human civilizations integrated concepts of prevention into their culture, their religion, and their laws. Prohibitions against specific foods—including pork, beef, and seafood—plus customs for food preparation, including officially-designated methods of killing cattle and methods of cooking, were part of the earliest practices of ancient societies. Prohibitions against alcohol or its limited use for religious ceremony have long been part of societies' efforts to control behavior, as well as prevent disease. Prohibition of cannibalism, the most

universal of food taboos, has strong grounding in the protection of health.[b]

Sexual practices have been viewed as having health consequences from the earliest civilizations. Male circumcision, premarital abstinence, and marital fidelity have all been shown to have impacts on health.

Quarantine or isolation of individuals with disease or exposed to disease has likewise been practiced for thousands of years. The intuitive notion that isolating individuals with disease could protect individuals and societies led to some of the earliest organized efforts to prevent the spread of disease. At times they were successful, but without a solid scientific basis. Efforts to separate individuals and communities from epidemics sometimes led to misguided efforts, such as the unsuccessful attempts to control the black plague by barring outsiders from walled towns and not recognizing that it was the rats and fleas that transmitted the disease.

During the 18th and first half of the 19th century individuals occasionally produced important insights into the prevention of disease. In the 1740s, British naval commander James Lind demonstrated that lemons and other citrus fruit could prevent and treat scurvy, a then-common disease of sailors whose daily nourishment was devoid of citrus fruit, the best source of vitamin C.

In the last years of the 18th century, English physician Edward Jenner recognized that cowpox, a common mild ailment of those who milked cows, protected those who developed it against life-threatening smallpox. He developed what came to be called a vaccine—derived from the Latin "*vacs*," meaning cows. He placed fluid from cowpox sores under the skin of recipients, including his son, and exposed them to smallpox. Despite the success of these smallpox prevention efforts, widespread use of vaccinations was slow to develop partially because at that time there was not an adequate scientific basis to explain the reason for its success.

All of these approaches to disease prevention were known before organized public health existed. Public health awareness began to emerge in Europe and America in the mid-19th century. The American public health movement had its origins in Europe where concepts of disease as the consequence of social conditions took root in the 1830s and 1840s. This movement, which put forth the idea that disease emerges from social conditions of inequality, produced the concept of

---

[a] Turnock[2] has described several meanings of public health. These include the system and social enterprise, the profession, the methods, the government services, and the health of the public. The population health approach used in this book may be thought of as subsuming all of these different perspectives on public health.

---

[b] In recent years, this prohibition has been indirectly violated by feeding beef products containing bones and brain matter to other cattle. The development of "mad cow" disease and its transmission to humans has been traced to this practice, which can be viewed as analogous to human cannibalism.

social justice. Many attribute public health's focus on vulnerable populations to this tradition.

While early organized public health efforts paid special attention to vulnerable members of society, they also focused on the hazards that affected everyone: contamination of the environment. This focus on sanitation and public health was often called the hygiene movement, which began even before the development of the germ theory of disease. Despite the absence of an adequate scientific foundation, the hygiene movement made major strides in controlling infectious diseases, such as tuberculosis, cholera, and waterborne diseases largely through alteration of the physical environment.

The fundamental concepts of epidemiology also developed during this era. In the 1850s, John Snow, often called the father of epidemiology, helped establish the importance of careful data collection and documentation of rates of disease before and after an intervention to evaluate effectiveness. He is known for his efforts to close down the Broad Street pump, which supplied water contaminated by cholera to a district of London. His actions quickly terminated that epidemic of cholera. John Snow's approach has become a symbol of the earliest epidemiological thinking.

Semmelweis, an Austrian physician, used much the same approach in the mid-19th century to control puerperal fever—or fever of childbirth—then a major cause of maternal mortality. Noting that physicians frequently went from autopsy room to delivery room without washing their hands, he instituted a hand washing procedure and was able to document a dramatic reduction in the frequency of puerperal fever. Unfortunately, he was unable to convince many of his contemporaries to accept this intervention without a clear mechanism of action. Until the acceptance of the germ theory of disease, puerperal fever continued to be the major cause of maternal deaths in Europe and North America.

The mid-19th century in England also saw the development of birth and death records, or vital statistics, which formed the basis of population-wide assessment of health status. From the beginning, there was controversy over how to define the cause of death. Two key figures in the early history of organized public health took opposing positions that reflect this continuing controversy. Edwin Chadwick argued that specific pathological conditions or diseases should be the basis for the cause of death. William Farr argued that underlying factors, including what we would today call risk factors and social conditions, should be seen as the actual causes of death.

Thus, the methods of public health were already being established before the development of the germ theory of disease by Louis Pasteur and his European colleagues in the mid-1800s. The revolutions in biology that they ignited ushered in a new era in public health. American physicians and public health leaders often went to Europe to study new techniques and approaches and brought them back to America to use at home.

After the Civil War, American public health began to produce its own advances and organizations. In 1872, the American Public Health Association (APHA) was formed. According to its own historical account, "the APHA's founders recognized that two of the association's most important functions were advocacy for adoption by the government of the most current scientific advances relevant to public health, and public education on how to improve community health."[8]

The biological revolution of the late 19th and early 20th centuries that resulted from the germ theory of disease laid the groundwork for the modern era of public health. An understanding of the contributions of bacteria and other organisms to disease produced novel diagnostic testing capabilities. For example, scientists could now identify tuberculosis cases through skin testing, bacterial culture, and the newly discovered chest X-ray. Concepts of vaccination advanced with the development of new vaccines against toxins produced by tetanus- and diphtheria-causing bacteria. Without antibiotics or other effective cures, much of public health in this era relied on prevention, isolation of those with disease, and case-finding methods to prevent further exposure.

In the early years of the 20th century, epidemiology methods continued to contribute to the understanding of disease. The investigations of pellagra by Goldberger and the United States Public Health Service overthrew the assumption of the day that pellagra was an infectious disease and established that it was a nutritional deficiency that could be prevented or easily cured with vitamin B-6 (niacin) or a balanced diet. Understanding of the role of nutrition was central to public health's emerging focus on prenatal care and childhood growth and development. Incorporating key scientific advances, these efforts matured in the 1920s and 1930s and introduced a growing alphabet of vitamins and nutrients to the American vocabulary.

A new public health era of effective intervention against active disease began in force after World War II. The discovery of penicillin and its often miraculous early successes convinced scientists, public health practitioners, and the general public that a new era in medicine and public health had arrived.

During this era, public health's focus was on filling the holes in the healthcare system. In this period, the role of public health was often seen as assisting clinicians to effectively deliver clinical services to those without the benefits of private medical care and helping to integrate preventive efforts into the practice of medicine. Thus, the great public health success of organized campaigns for the eradication of polio was mistakenly seen solely as a victory for medicine. Likewise, the successful passage of Medicaid and Medicare, outgrowths of public health's commitment to social justice, was simply viewed as efforts to expand the private practice of medicine.

This period, however, did lay the foundations for the emergence of a new era in public health. Epidemiological methods designed for the study of noncommunicable diseases demonstrated the major role that cigarette smoking plays in lung cancer and a variety of other diseases. The emergence of the randomized clinical trial and the regulation of drugs, vaccines, and other interventions by the Food and Drug Administration developed the foundations for what we now call evidence-based public health and evidence-based medicine.

The 1980s and much of the 1990s were characterized by a focus on individual responsibility for health and interventions at the individual level. Often referred to as health promotion and disease prevention, these interventions targeted individuals to effect behavioral change and combat the risk factors for diseases. As an example, to help prevent coronary artery disease, efforts were made to help individuals address high blood pressure and cholesterol, cigarette smoking, and obesity. Behavioral change strategies were also used to help prevent the spread of the newly emerging HIV/AIDS epidemic. Efforts aimed at individual prevention and early detection as part of medical practice began to bear some fruit with the widespread introduction of mammography for detection of breast cancer and the worldwide use of Pap smears for the detection of cervical cancer. Newborn screening for genetic disease became a widespread and often legally-mandated program, combining individual and community components.

Major public health advances during this era resulted from the environmental movement, which brought public awareness to the health dangers of lead in gasoline and paint. The environmental movement also focused on reducing cancer by controlling radiation exposure from a range of sources including sunlight and radon, both naturally-occurring radiation sources. In a triumph of global cooperation, governments worked together to address the newly-discovered hole in the ozone layer. In the United States, reductions in air pollution levels and smoking rates during this era had an impact on the frequency of chronic lung disease, asthma, and most likely coronary artery disease.

The heavy reliance on individual interventions that characterized much of the last half of the 20th century changed rapidly in the beginning of the 21st century. A new era in public health that is often called "population health" has begun to transform professional and public thought about health. From the potential for bioterrorism to the high costs of health care to the control of pandemic influenza and AIDS, the need for

community-wide or population-wide, public health efforts have become increasingly evident. This new era is characterized by a global perspective and the need to address international health issues. It includes a focus on the potential impacts of climate change, emerging and reemerging infectious diseases, and the consequences of trade in potentially contaminated or dangerous products, ranging from food to toys.

Table 1-1 outlines these eras of public health, identifies their key defining elements, and highlights important events that symbolize each era.[9]

Thus, today we have entered an era in which a focus on the individual is increasingly coupled with a focus on what needs to be done at the community and population level. This era of public health can be viewed as "the era of population health."

WHAT IS MEANT BY POPULATION HEALTH?

The concept of population health has emerged in recent years as a broader concept of public health that includes all the ways that society as a whole or communities within society are affected by health issues and how they respond to these issues. Population health provides an intellectual umbrella for thinking about the wide spectrum of factors that can and do affect the health of individuals and the population as a whole. Figure 1-1 provides an overview of what falls under the umbrella of population health.

Population health also provides strategies for considering the broad range of potential **interventions** to address these issues. By intervention we mean the full range of strategies designed to protect health, and prevent disease, disability, and death. Interventions include: preventive efforts, such as nutrition and vaccination; curative efforts, such as antibiotics and cancer surgery; and efforts to prevent complications and restore function—from chemotherapy to physical therapy. Thus, population health is about *healthy people and healthy populations*.

The concept of population health can be seen as a comprehensive way of thinking about the modern scope of public health. It utilizes an evidence-based approach to analyze the determinants of health and disease and the options for intervention to preserve and improve health. Population health requires us to define what we mean by **health issues** and what we mean by **population(s)**. It also requires us to define what we mean by **society's shared health concerns**, as well as **society's vulnerable groups**.

To understand population health, we therefore need to define what we mean by each of these four components:

- Health issues
- Population(s)
- Society's shared health concerns
- Society's vulnerable groups

TABLE 1-1 Eras of Public Health

Eras of public health	Focus of attention/ Paradigm	Action framework	Notable events and movements in public health and epidemiology
Health protection (Antiquity–1830s)	Authority-based control of individual and community behaviors	Religious and cultural practices and prohibited behaviors	Quarantine for epidemics; sexual prohibitions to reduce disease transmission; dietary restrictions to reduce food-borne disease
Hygiene movement (1840–1870s)	Sanitary conditions as basis for improved health	Environmental action on a community-wide basis distinct from health care	Snow on Cholera; Semmelweis and puerperal fever; collection of vital statistics as empirical foundation for public health and epidemiology
Contagion control (1880–1940s)	Germ theory: demonstration of infectious origins of disease	Communicable disease control through environmental control, vaccination, sanatoriums, and outbreak investigation in general population	Linkage of epidemiology, bacteriology, and immunology to form TB sanatoriums; outbreak investigation, e.g., Goldberger and pellagra
Filling holes in the medical care system (1950s–mid-1980s)	Integration of control of communicable diseases; modification of risk factors; and care of high-risk population as part of medical care	Public system for care of and control of specific infectious diseases and vulnerable populations distinct from general health care system; Integrated health maintenance organizations with integration of preventive services into general health care system	Antibiotics; randomized clinical trials; concept of risk factors; Surgeon General reports on cigarette smoking; Framingham study on cardiovascular risks; health maintenance organizations and community health centers with integration of preventive services into general healthcare system
Health promotion/ Disease prevention (Mid-1980–2000)	Focus on individual behavior and disease detection in vulnerable and general populations	Clinical and population-oriented prevention with focus on individual control of decision making and multiple interventions	AIDS epidemic and need for multiple interventions to reduce risk; reductions in coronary heart disease through multiple interventions
Population health (21st century)	Coordination of public health and health care delivery based upon shared evidence-based systems thinking	Evidence-based recommendations and information management; focus on harms and costs as well as benefits of interventions; globalization	Evidence-based medicine and public health; information technology; medical errors; antibiotic resistance; global collaboration, e.g., SARS, tobacco control, climate change

Source: Awofeso N. What's new about the "New Public Health"? *American Journal of Public Health.* 2004;94(5):705–709.

FIGURE 1-1 The full spectrum of population health

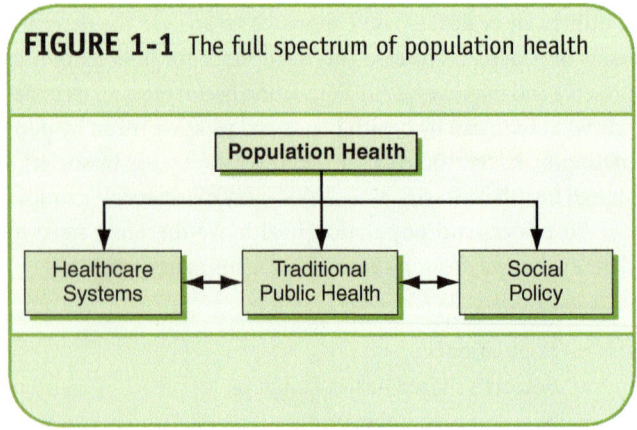

What Are the Implications of Each of the Four Components of Population Health?

All four of the key components of public health have changed in recent years. Let us take a look at the historical, current, and emerging scopes of each component and consider their implications.

For most of the history of public health, the term "health" focused solely on physical health. Mental health has now been recognized as an important part of the definition; conditions such as depression and substance abuse make enormous contributions to disability in populations throughout the world. The boundaries of what we mean by health continue to expand and the limits of health are not clear. Many novel medical interventions, including modification of genes and treatments to increase height, improve cosmetic appearance, and improve sexual performance, confront us with the question: are these health issues?

The definition of a population, likewise, is undergoing fundamental change. For most of recorded history, a population was defined geographically. Geographic communities, such as cities, states, and countries, defined the structure and functions of public health. The current definition of population has expanded to include the idea of a global community, recognizing the increasingly interconnected issues of global health. The definition of population is also focusing more on nongeographic communities. Universities now include the distance-learning community; health care is delivered to members of a health plan; and the Internet is creating new social communities. All of these new definitions of a population are affecting the thinking and approaches needed to address public health issues.

What about the meaning of society-wide concerns—have they changed as well? Historically, public health and communicable disease were nearly synonymous, as symbolized by the field of epidemiology which actually derives its name from the study of communicable disease epidemics. In recent decades, the focus of society-wide concerns has greatly expanded to include toxic exposures from the physical environment, transportation safety, and the costs of health care. However, communicable disease never went away as a focus of public health and the 21st century is seeing a resurgence in concern over emerging infectious diseases, including HIV/AIDS, pandemic flu, and newly drug-resistant diseases, such as staph infections and tuberculosis. Additional concerns, ranging from the impact of climate change to the harms and benefits of new technologies, are altering the meaning of society-wide concerns.

Finally, the meaning of vulnerable populations continues to transform. For most of the 20th century, public health focused on maternal and child health and high risk occupations as the operational definition of vulnerable populations. While these groups remain important to public health, additional groups now receive more attention, including the disabled, the frail elderly, and those without health insurance. Attention is also beginning to focus on the immune-suppressed among those living with HIV/AIDS, who are at higher risk of infection and illness, and those whose genetic code documents their special vulnerability to disease.

Public health has always been about our shared health concerns as a society and our concerns about vulnerable populations. These concerns have changed over time, and new concerns continue to emerge. Table 1-2 outlines historical, current, and emerging components of the population health approach to public health. As is illustrated by communicable diseases, past concerns cannot be relegated to history.

SHOULD WE FOCUS ON EVERYONE OR ON VULNERABLE GROUPS?

Public health is often confronted with the potential conflict of focusing on everyone and addressing society-wide concerns

TABLE 1-2 Components of Population Health

	Health	Population	Examples of society-wide concerns	Examples of vulnerable groups
Historical	Physical	Geographically limited	Communicable disease	High risk maternal and child, high risk occupations
Current	Physical and mental	Local, state, national, global, governmentally-defined	Toxic substances, product and transportation safety, communicable diseases, costs of health care	Disabled, frail elderly, uninsured
Emerging	Cosmetic, genetic, social functioning	Defined by local, national, and global communications	Disasters, climate change, technology hazards, emerging infectious diseases	Immune-suppressed, genetic vulnerability

versus focusing on the needs of vulnerable populations.[10] This conflict is reflected in the two different approaches to addressing public health problems. We will call them the **high-risk approach** and the **improving-the-average approach**.

The high-risk approach focuses on those with the highest probability of developing the disease and aims to bring their risk close to the levels experienced by the rest of the population. Figure 1-2A illustrates the high risk approach.

The success of the high-risk approach, as shown in Figure 1-2B, assumes that those with a high probability of developing disease are heavily concentrated among those with exposure to what we call **risk factors**. Risk factors include a wide range of exposures from cigarette smoke and other toxic substances to high risk sexual behaviors.

The improving-the-average approach focuses on the entire population and aims to reduce the risk for everyone. Figure 1-3 illustrates this approach.

The improving-the-average approach assumes that everyone is at some degree of risk and the risk increases with the extent of exposure. In this situation, most of the disease occurs

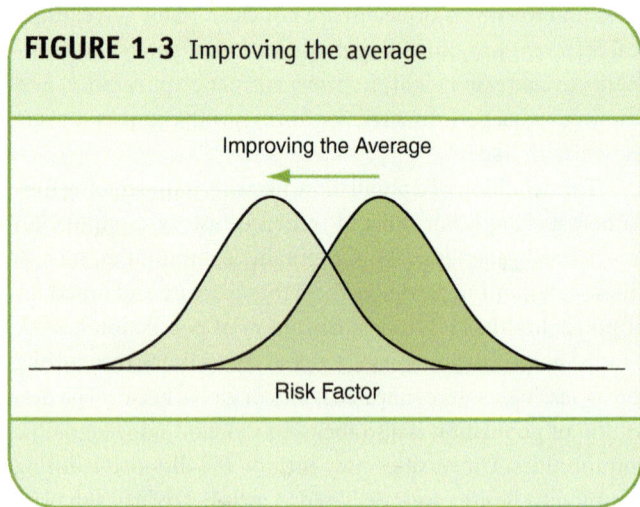

FIGURE 1-3 Improving the average

among the large number of people who have only modestly increased exposure. The successful reduction in average cholesterol levels through changes in the American diet and the anticipated reduction in diabetes via a focus on weight reduction among children illustrate this approach.

One approach may work better than the other in specific circumstances, but in general both approaches are needed if we are going to successfully address today's and tomorrow's health issues. These two approaches parallel public health's long-standing focus on both the health of vulnerable populations and society-wide health concerns.[c]

Now that we understand what is meant by population health, let us take a look at the range of approaches that may be used to promote and protect health.

WHAT ARE THE APPROACHES AVAILABLE TO PROTECT AND PROMOTE HEALTH?

The wide range of strategies that have been, are being, and will be used to address health issues can be divided into three general categories: health care, traditional public health, and social interventions.

Health care includes the delivery of services to individuals on a one-on-one basis. It includes services for those who are sick or disabled with illness or diseases, as well as for those who are asymptomatic. Services delivered as part of

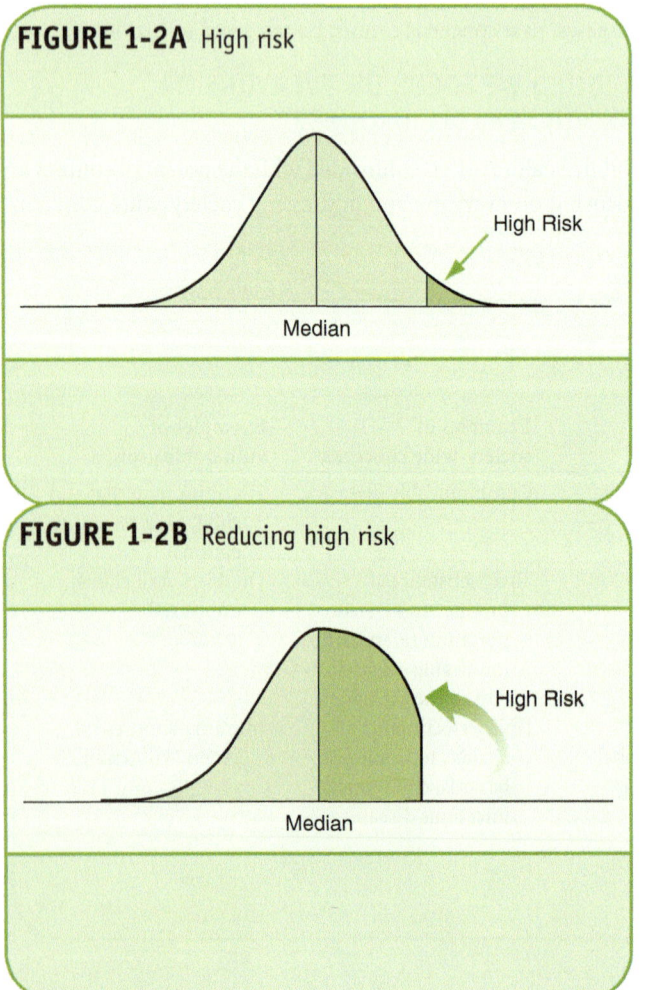

FIGURE 1-2A High risk

FIGURE 1-2B Reducing high risk

---

[c] An additional approach includes reducing disparities by narrowing the curve. For instance, this might be accomplished by transferring financial and/or health services from the low risk to the high risk category through taxation or other methods. Depending on the distribution of the factors affecting health, this approach may or may not reduce the overall frequency of disease more than the other approaches. The distribution of risk in Figures 1-1 and 1-2 assumes a bell-shaped or normal distribution. The actual distribution of factors affecting health may not follow this distribution.

clinical prevention have been categorized as vaccinations, behavioral counseling, screening for disease, and preventive medications.[11]

Traditional public health efforts have a population-based preventive perspective utilizing interventions targeting communities or populations, as well as defined high risk or vulnerable groups. Communicable disease control, reduction of environmental hazards, food and drug safety, and nutritional and behavioral risk factors have been key areas of focus of traditional public health approaches.

Both health care and traditional public health approaches share a goal to directly affect the health of those they reach. In contrast, social interventions are primarily aimed at achieving other nonhealth goals, such as increasing convenience, pleasure, economic growth, and social justice. Social interventions range from improving housing, improving education and services for the poor, to increased global trade. These interventions may have dramatic and sometimes unanticipated positive or negative health consequences. Social interventions, like increased availability of food, may improve health, while the availability of high-fat or high-calorie foods may pose a risk to health.

Table 1-3 describes the characteristics of health care, traditional public health, and social approaches to population health and provides examples of each approach.

None of these approaches is new. However, they have traditionally been separated or put into silos in our thinking process with the connections between them often ignored. Connecting the pieces is an important part of the 21st century challenge of defining public health.

Now that we have explained what we mean by public health and seen the scope and methods that we call population health, let us continue our big-picture approach by taking a look at what we mean by the determinants of health and disease.

WHAT FACTORS DETERMINE THE OCCURRENCE OF DISEASE, DISABILITY, AND DEATH?

To complete our look at the big picture issues in public health, we need to gain an understanding of the forces that determine disease and the outcome of disease including what in public health has been called morbidity (disability) and mortality (death).[d]

We need to establish what are called contributory causes based on evidence. **Contributory causes** can be thought of as causes of disease. For instance, the HIV virus and cigarette smoking are two well-established contributory causes of disease, disability, and death. They produce disease, as well as disability and death. However, knowing these contributory causes of disease is often not enough. We need to ask: what determines whether people will smoke or come in contact with the HIV virus? What determines their course once exposed to cigarettes or HIV? In public health we use the term **determinants** to identify these underlying factors that ultimately bring about disease.

Determinants look beyond the known contributory causes of disease to factors that are at work often years before a

---

[d] We will use the term "disease" as shorthand for the broad range of outcomes that includes injuries and exposures that result in death and disability.

TABLE 1-3 Approaches to Population Health

	Characteristics	Examples
Health care	Systems for delivering one-on-one individual health services including those aimed at prevention, cure, palliation, and rehabilitation	Clinical preventive services including: vaccinations, behavioral counseling, screening for disease, and preventive medications
Traditional public health	Group- and community-based interventions directed at health promotion and disease prevention	Communicable disease control, control of environmental hazards, food and drug safety, reduction in risk factors for disease
Social	Interventions with another nonhealth-related purpose, which have secondary impacts on health	Interventions that improve the built environment, increase education, alter nutrition, or address socioeconomic disparities through changes in tax laws; globalization and mobility of goods and populations

disease develops.[12,13] These underlying factors may be thought of as "upstream" forces. Like great storms, we know the water will flow downstream, often producing flooding and destruction along the way. We just don't know exactly when and where the destruction will occur.

There is no official list or agreed-upon definition of what is included in determinants of disease.[e] Nonetheless, there is wide agreement that the following factors are among those that can be described as determinants in that they increase or at times decrease the chances of developing conditions that threaten the quantity and/or quality of life.

Behavior
Infections
Genetics

Geography
Environment
Medical care
Socio-economic-cultural

BIG GEMS provides a convenient device for remembering these determinants of disease. Let's see what we mean by each of the determinants.

Behavior—Behavior implies actions that increase exposure to the factors that produce disease or protect individuals from disease. Actions such as cigarette smoking, exercise, diet, alcohol consumption, unprotected intercourse, and seat belt use are all examples of the ways that behaviors help determine the development of disease.

Infection—Infections are often the direct cause of disease. In addition, we are increasingly recognizing that early or long-standing exposures to infections may contribute to the development of disease or even protection against disease. Diseases as diverse as gastric and duodenal ulcers, gallstones, and hepatoma or cancer originating in the liver, are increasingly suspected to have infection as an important determinant of the disease. Early exposure to infections may actually reduce diseases ranging from polio to asthma.

Genetics—The revolution in genetics has focused our attention on roles that genetic factors play in the development

and outcome of disease. Even when contributory causes, such as cigarettes, have been clearly established as producing lung cancer, genetic factors also play a role in the development and progression of the disease. While genetic factors play a role in many diseases, they are only occasionally the most important determinant of disease.

Geography—Geographic location influences the frequency and even the presence of disease. Infectious diseases such as malaria, Chagas disease, schistosomiasis, and Lyme disease occur only in defined geographic areas. Geography may also imply local geological conditions, such as those that produce high levels of radon—a naturally-occurring radiation that contributes to the development of lung cancer.

Environment—Environmental factors determine disease and the course of disease in a number of ways. The unaltered or "natural" physical world around us may produce disability and death from sudden natural disasters, such as earthquakes and volcanic eruptions, to iodine deficiencies due to low iodine content in the food-producing soil. The altered physical environment produced by human intervention includes exposures to toxic substances in occupational or nonoccupational settings. The physical environment built for use by humans—the **built environment**—produces determinants ranging from indoor air pollution, to "infant-proofed" homes, to hazards on the highway.

Medical care—Access to and the quality of medical care can be a determinant of disease. When a high percentage of individuals are protected by vaccination, nonvaccinated individuals in the population may be protected as well. Cigarette smoking cessation efforts may help smokers to quit, and treatment of infectious disease may reduce the spread to others. Medical care, however, often has its major impact on the course of disease by attempting to prevent or minimize the disability and death once disease develops.

Social-economic-cultural—In the United States, socioeconomic factors have been defined as education, income, and occupational status. These measures have all been shown to be determinants of diseases as varied as breast cancer, tuberculosis, and occupational injuries. Cultural and religious factors are increasingly being recognized as determinants of diseases because beliefs sometimes influence decisions about treatments, in turn affecting the outcome of the disease. While most diseases are more frequent in lower socioeconomic groups, others such as breast cancer are often more common in higher socioeconomic groups.

We will return to determinants again and again as we explore the work of population health. Historically, understanding determinants has often allowed us to prevent diseases and their consequences even when we did not fully understand the

---

[e] Health Canada[12] has identified 12 determinants of health that are: 1) income and social status; 2) employment; 3) education; 4) social environments; 5) physical environments; 6) healthy child development; 7) personal health practices and coping skills; 8) health services; 9) social support networks; 10) biology and genetic endowment; 11) gender; and 12) culture. Many of these are subsumed under socio-economic-cultural determinants in the BIG GEMS framework. The World Health Organization's Commission on Social Determinants of Health has also produced a list of determinants that is consistent with the BIG GEMS framework.[13]

mechanism by which the determinants produced their impact. For instance:

- Scurvy was controlled by citrus fruits well before vitamin C was identified.
- Malaria was partially controlled by clearing swamps before the relationship to mosquito transmission was appreciated.
- Hepatitis B and HIV infections were partially controlled even before the organisms were identified through reduction in use of contaminated needles and blood transfusions.
- Tuberculosis death rates were greatly reduced through less crowded housing, the use of TB sanitariums, and better nutrition.

Using asthma as an example, Box 1-1 illustrates the many ways that determinants can affect the development and course of a disease.

Thus, population health focuses on the big picture issues and the determinants of disease. Increasingly, public health also emphasizes a focus on the research evidence as a basis for understanding the cause or etiology of disease and the intervention that can improve the outcome.

BOX 1-1 Asthma and the Determinants of Disease.

Jennifer, a teenager living in an urban rundown apartment in a city with high levels of air pollution, develops severe asthma. Her mother also has severe asthma, yet both of them smoke cigarettes. Her clinician prescribed medications to prevent asthma attacks, but she takes them only when she experiences severe symptoms. Jennifer is hospitalized twice with pneumonia due to common bacterial infections. She then develops an antibiotic-resistant infection. During this hospitalization, she requires intensive care on a respirator. After several weeks of intensive care and every known treatment to save her life, she dies suddenly.

Asthma is an inflammatory disease of the lung coupled with an increased reactivity of the airways, which together produce a narrowing of the airways of the lungs. When the airways become swollen and inflamed, they become narrower, allowing less air through to the lung tissue and causing symptoms such as wheezing, coughing, chest tightness, breathing difficulty, and predisposition to infection. Once considered a minor ailment, asthma is now the most common chronic disorder of childhood. It affects over six million children under the age of 18 in the United States alone.

Jennifer's tragic history illustrates how a wide range of determinants of disease may affect the occurrence, severity, and development of complications of a disease. Let's walk through the BIG GEMS framework and see how each determinant impacts in Jennifer's story.

Behavior—Behavioral factors play an important role in the development of asthma attacks and in their complications. Cigarette smoking makes asthma attacks more frequent and more severe. It also predisposes individuals to developing infections such as pneumonia. Treatment for severe asthma requires regular treatments along with more intensive treatment when an attack occurs. It is difficult for many people, especially teenagers, to take medication regularly, yet failure to adhere to treatment greatly complicates the disease.

Infection—Infection is a frequent precipitant of asthma and asthma increases the frequency and severity of infections. Infectious diseases, especially pneumonia, can be life-threatening in asthmatics requiring prompt and high quality medical care. The increasing development of antibiotic-resistant infections pose special risks to those with asthma.

Genetics—Genetic factors predispose people to childhood asthma. However, many children and adults without a family history develop asthma.

Geography—Asthma is more common in geographic areas with high levels of naturally occurring allergens due to flowering plants. However, today even populations in desert climates in the United States are often affected by asthma, as irrigation results in the planting of allergen-producing trees and other plants.

Environment—The physical environment, including that built for use by humans, has increasingly been recognized as a major factor affecting the development of asthma and asthma attacks. Indoor air pollution is the most common form of air pollution in many developing countries. Along with cigarette smoke, air pollution inflames the lungs acutely and chronically. Cockroaches often found in rundown buildings have been found to be highly allergenic and predisposing to asthma. Other factors in the built environment, including mold and exposure to pet dander, can also trigger wheezing in susceptible individuals.

(continues)

BOX 1-1 *continued.*

Medical care—The course of asthma can be greatly affected by medical care. Management of the acute and chronic effects of asthma can be positively affected by efforts to understand an individual's exposures, reducing the chronic inflammation with medications, managing the acute symptoms, and avoiding life-threatening complications.

Socio-economic-cultural—Disease and disease progression are often influenced by an individual's socioeconomic status. Air pollution is often greater in lower socioeconomic neighborhoods of urban areas. Mold and cockroach infestations may be greater in poor neighborhoods. Access to and quality of medical care may be affected by social, economic, and cultural factors.

Thus, asthma is a condition which demonstrates the contributions made by the full range of determinants included in the BIG GEMS framework. No one determinant alone explains the bulk of the disease. The large number of determinants and their interactions provide opportunities for a range of health care, traditional public health, and social interventions.

Key Words

- Population health approach
- Social justice
- Interventions
- Health issues
- Population(s)
- Society's shared health concerns
- Society's vulnerable groups
- High-risk approach
- Improving-the-average approach
- Risk factor
- Contributory causes
- Determinants
- Built environment
- Behavior
- Infections
- Genetics
- Geography
- Environment
- Medical care
- Socio-economic-cultural

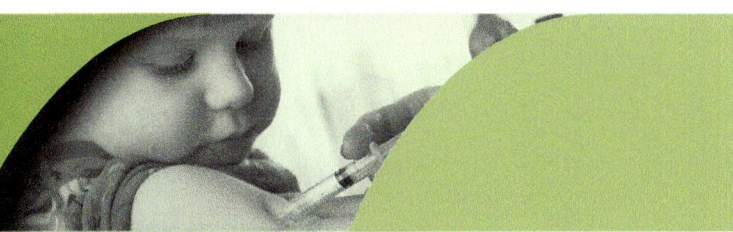

Discussion Question

1. Think about a typical day in your life and identify ways that public health affects it.

REFERENCES

1. Pfizer Global Pharmaceuticals. *Milestones in Public Health: Accomplishments in Public Health over the Last 100 Years.* New York: Pfizer Global Pharmaceuticals; 2006.

2. Turnock BJ. *Public Health: What It Is and How It Works,* 4th ed. Sudbury, MA: Jones and Bartlett Publishers; 2009.

3. Winslow CEA. The untilled field of public health. *Mod. Med.* 1920; 920;2:183–191.

4. Institute of Medicine. *The Future of Public Health.* Washington, DC: National Academy Press; 1988:41.

5. Young TK. *Population Health: Concepts and Methods.* New York: Oxford University Press; 1998.

6. Rosen G. *A History of Public Health.* Baltimore: Johns Hopkins University Press; 1993.

7. Porter D. *Health, Civilization, and the State: A History of Public Health from Ancient to Modern Times.* Oxford: Rutledge; 1999.

8. American Public Health Association. APHA History and Timeline. Available at: http://www.apha.org/about/news/presskit/aphahistory.htm? NRMODE=Published&NRNODEGUID=%7b8AF0A3FE-8B29-4952-87EF-2757B9B2668F%7d&NRORIGINALURL=%2fabout%2fnews%2fpresskit% 2faphahistory.htm&NRCACHEHINT=NoModifyGuest&PF=true. Accessed March 12, 2009.

9. Awofeso N. What's new about the "New Public Health"? *American Journal of Public Health.* 2004;94(5):705–709.

10. Rose G, Khaw KT, Marmot M. *Rose's Strategy of Preventive Medicine.* New York: Oxford University Press; 2008.

11. Agency for Healthcare Research and Quality. Preventive Services. Available at: http://www.ahrq.gov/. Accessed March 12, 2009.

12. Public Health Agency of Canada. Population Health Approach— What Determines Health? Available at: http://www.phac-aspc.gc.ca/ph-sp/ determinants/index-eng.php. Accessed March 12, 2009.

13. Commission on Social Determinants of Health. *Closing the gap in a generation: health equity through action on the social determinants of health. Final Report of the Commission on Social Determinants of Health.* Geneva: World Health Organization; 2008.

The Immune Response

The immune system is complex; if it was simple, it would be practically useless.

—Unknown

Preview

The word "immunity," in its broadest sense, means freedom from a burden, be it legal action, taxes, or, in the present context, disease. In the first season of the hit television show "Survivor," Richard Hatch wore the "immunity necklace," which meant that his fellow castaways could not vote him off the island. Witnesses who may have been involved in criminal activity can be granted immunity from prosecution so they may testify against those involved in more serious crimes. In the early years of the colonization of the United States, the battle cry of the colonists was "immunity from taxes."

The ultimate outcome of infection is the result of the dynamic interplay between two opposing forces summarized by the expression $D = nV/R$. In the equation D refers to microbial disease, the numerator refers to the microbe and its virulence factors, and the denominator (R) is a function of the host immune system. The interplay determines whether no infection, mild infection, or severe and, possibly, fatal infection will result. The early hours of encounter between the microbe and the immune system are particularly crucial in determining the eventual outcome.

The immune system functions to recognize and destroy foreignness as embodied in invading microbes and their products and in mutant, damaged, and worn out cells. This is accomplished by both nonspecific (inherent) immunity and specific (acquired) immunity, functions accomplished by a variety of body systems working together. Nonspecific immunity is characterized by physiological defenses that act to prevent microbes from gaining access into the body and facilitate the elimination of those that have penetrated. Specific immunity targets microbes for elimination, based on the recognition of their foreign antigens, by antibody-mediated and cell-mediated immunity. Impairment of the immune system as occurs in AIDS, leukemia, immunosuppressive drugs, or congenital disease renders the immuno-compromised individual subject to repeated life-threatening infections.

AUTHOR'S NOTE
In writing this text I was initially in a dilemma. Should the microbes and their virulence mechanisms precede the body's immune defense, or the other way around? It is the "what came first, the chicken or the egg?" puzzle. Because a strong focus in this text is on disease prevention, it seemed more logical to first present what it is that the immune system is combating. I hope you agree!

■ Basic Concepts

The immune system functions to recognize and destroy that which is **foreign**, including invading microbes and their secretions, toxins and enzymes for example, and "non-normal" self cells. Functionally, the immune system consists of two components: **nonspecific** or innate (inherent) immunity and **specific** or adapted (acquired after birth) immunity (FIGURE 2.1). Both mechanisms are the strategies by which the body eliminates "foreignness."

There is no one good way to measure the effectiveness of the immune system because of its many complex interactions. In fact, ability to cope with the constant

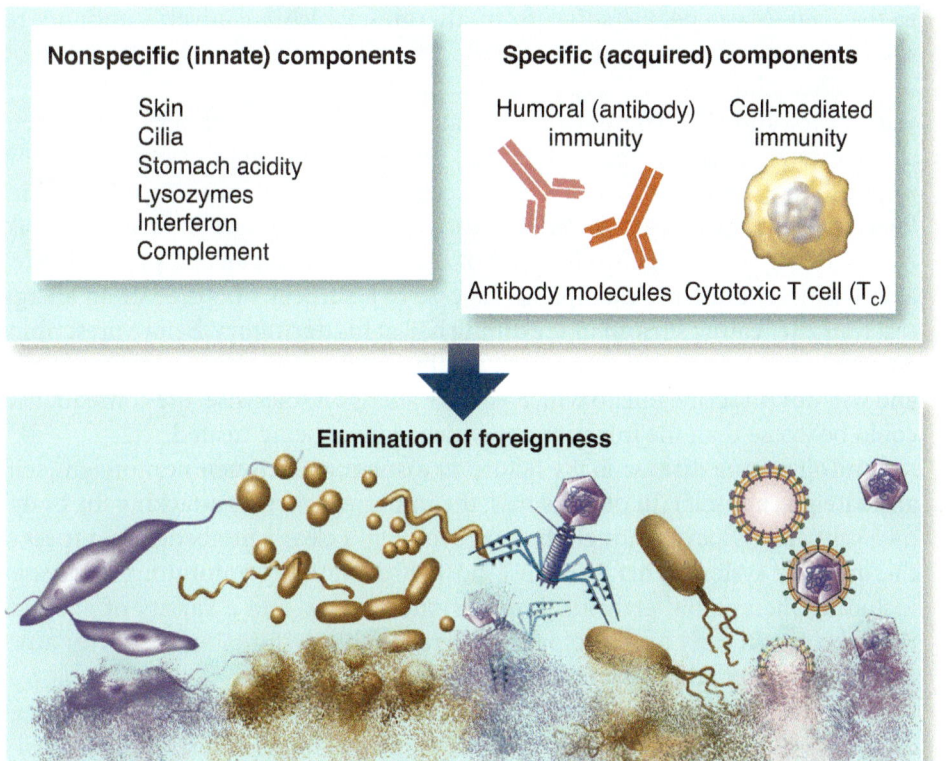

FIGURE 2.1 Nonspecific and specific components of the immune system protect against foreignness.

barrage of microorganisms is not constant, nor is it measurable. Age, sex, nutrition, and healthiness, along with physical and mental stress, play a significant role in immune status. Think about final exam time. Lights are on late in dormitories and in study halls around the campus, and fast foods are consumed at an even more rapid rate than usual as you review a semester's work for four or five courses in about a week's time. There is no doubt that your immune system is temporarily compromised, making you more vulnerable to infection.

Immune status varies within the individual and between individuals. Consider, for example, that a hundred people may be crowded into a lecture hall, movie theater, or restaurant and that all are exposed to circulating cold viruses. Some will "catch the cold," and others will not. Obviously, the numerator in the equation $D = nV/R$ is the same for all who are exposed, and, therefore, whether infection results is largely a function of R, the immune status at that time. A recent news article cited two deaths that occurred from babesiosis, a tickborne protozoan disease that is rarely fatal unless the individual has an underlying health problem. This was the case in both patients who died. Genetics, too, plays a key role; it is well established that there are ethnic and racial differences in susceptibility to microbial infection.

Although the primary role of the immune system is as a mechanism of defense against microbial infection, it also acts as a system of internal surveillance leading to elimination of tumor cells and destruction of old, worn out, and damaged red blood cells. These cells, like microbes, are recognized as foreign and trigger an immune response.

The immune system is beneficial for the most part, but there is another side to the coin. Many people suffer from **allergies**, which are adverse immune responses to protein molecules associated with pollens, dust, foods, mites, antibiotics, and bee stings. These allergies run the gamut from being a nuisance (sneezing, watery eyes, and runny nose) to life-threatening anaphylactic shock. Anaphylactic shock occurs when the body releases an overwhelming amount of histamine in response to the allergen. The histamine causes the blood vessels to dilate, which lowers blood pressure, and severely constricts the bronchioles in the lungs, making breathing very difficult. The truth of the matter is that an allergy can develop at any time. A dramatic example is the case of a student who died of an allergic reaction after eating shrimp in the dining hall at his dormitory. Before prescribing an antibiotic, a physician should ask whether you are allergic to any antibiotics and will not prescribe one to which you are allergic. Otherwise, the consequences could be worse than the infection for which you are being treated.

Autoimmune disease is the failure to distinguish between nonforeign (self) and foreign (nonself). In other words, the immune system is attacking the body's own cells and tissues as though they were foreign, a clearly nonbeneficial aspect of the immune system. This leads to a host of debilitating autoimmune diseases, including rheumatoid arthritis, lupus erythematosus, and a variety of anemias. Paul Ehrlich, an early immunologist, referred to this dysfunction as "horror autotoxicus," a fear of self-poisoning.

The major obstacle to transplantation of organs is rejection of transplanted organs by the immune systems of their recipients. Should this be considered a nonbeneficial aspect of the immune system? Most immunologists would not think so, because, after all, the recipient is responding to foreign molecules of the donor cells in a manner nature evolved to thwart microbial invasion and to conduct internal surveillance. Transplantation technology is not a natural phenomenon but is the result of advanced medical science.

Anatomy and Physiology of the Body's Defenses

Having now established a few basic concepts involving the immune system, attention needs to be focused on anatomical and physiological considerations. Perhaps in a precollege course in biology you dissected a frog; recall that the digestive system is a series of defined anatomical structures starting with the oral cavity, which led to the esophagus, which opened into the stomach, and so forth. The same organization can be said for all the systems of the body. There is a hierarchy of cells, tissues, and organs in biological systems. The immune system is, however, unique in that the anatomical structures are, for the most part, shared with more defined systems. The spleen, in most anatomy manuals, is described with the circulatory system, and the lymph nodes are considered components of the lymphatic system, but these structures are also parts of the immune system. The major structures of the immune system are shown in FIGURE 2.2.

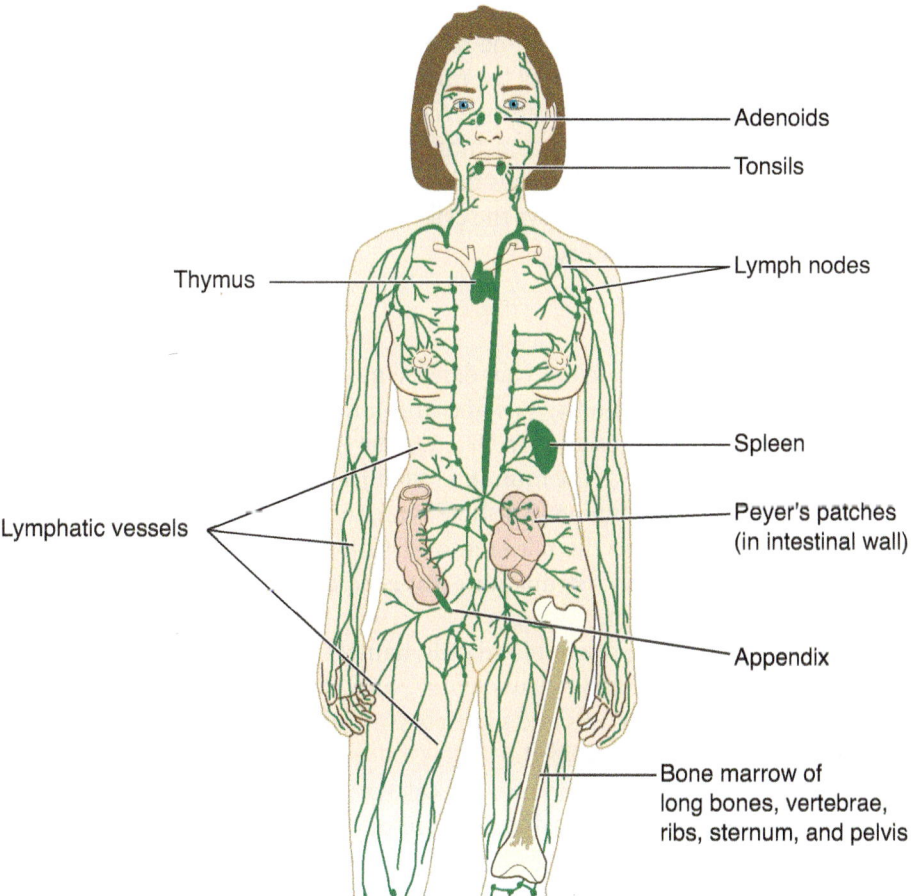

FIGURE 2.2 Anatomy of the immune system.

Adenoids

Tonsils

Lymph nodes

Thymus

Spleen

Peyer's patches (in intestinal wall)

Lymphatic vessels

Appendix

Bone marrow of long bones, vertebrae, ribs, sternum, and pelvis

Mechanical and Chemical Barriers

The skin is the first line of defense against microbial infection. It is a physical barrier that, when intact, blocks the entry of microorganisms into the body based on its structural composition. Obviously, there are circumstances in which the skin is broken, allowing possible penetration by microbes. Cuts and abrasions, insect bites, and injections with hypodermic needles are familiar examples. Staphylococci, normal residents on the surface of the skin, are particularly prone to enter the body via these breaks and establish infection. Many diseases, including malaria, yellow fever, babesiosis, and sleeping sickness (other than in the classroom), are transmitted by insects whose bites penetrate the skin.

The acidity of the stomach is detrimental to most microbes. Certainly, food and eating utensils are not sterile, and, hence, large numbers of microbes gain access to the stomach. In addition, mucous membranes line the internal body cavities and act as internal barriers. For example, the digestive tract running from the mouth to the anus is lined by a mucous membrane that protects the body from invasion by microbes that reside within the digestive tube. Penetration of this mucous membrane by intestinal microbes, even those that normally reside within the digestive tract, is extremely serious and leads to a life-threatening infection known as **peritonitis**. Surgeons are very much aware of this when performing procedures on the digestive system; nicking the lining of the intestinal tract allows microorganisms to spill out into body cavities where they do not belong.

The airways in certain areas of the respiratory system have cells with hairlike projections, called **cilia**, that propel the microbes into the pharynx where they are swallowed and expelled with fecal material. This system is called the mucosal-ciliary escalator system. Smokers are more prone to respiratory infections than nonsmokers, because smoke damages this system.

The enzyme **lysozyme** destroys the cell walls of some bacteria. Lysozymes are found in saliva, tears, and sweat.

Blood

Several components of blood are vital defense mechanisms of the body and contribute to nonspecific immunity, specific immunity, or both. A summary of the components of blood is presented in TABLE 2.1. The globulin fraction is particularly important to the function of the immune system; a subfraction, **gamma globulin**, contains most of the **antibodies**. Antibodies are molecules produced in response to antigens and act to target them for removal. Conversely, **antigens** are molecules that bring about antibody production. Antigens and antibodies are described more fully in a later section.

There are three categories of blood cells, namely the **red blood cells** (erythrocytes), the **white blood cells** (leukocytes), and the **platelets** (thrombocytes) (FIGURE. 2.3). The red blood cells and platelets play no direct role in immunity and therefore are not further described.

White blood cells are divided into five categories (TABLE 2.2). Neutrophils, basophils, and eosinophils are referred to as granulocytes because microscopic examination reveals the presence of granules in their cytoplasm. Lymphocytes and monocytes do not contain cytoplasmic granules, and these white blood cells are

TABLE 2.1 Composition of Blood

Component	Function
Plasma	
Water (80–90%)	Solvent
Proteins	
Albumin	Blood volume
Globulins	Antibody molecules
Fibrinogen	Blood clotting
Complement	Immune amplification
Interferon	Antiviral properties
Blood cells[a]	
Red blood cells (erythrocytes)	Oxygen and carbon dioxide transport
White blood cells (leukocytes)	Antibody formation, CMI, phagocytosis
Thrombocytes (platelets)	Blood clotting

[a]In each microliter of blood there are about 4.5 to 5.5 million erythrocytes, 5,000 to 10,000 leukocytes, and 250,000 to 400,000 thrombocytes.

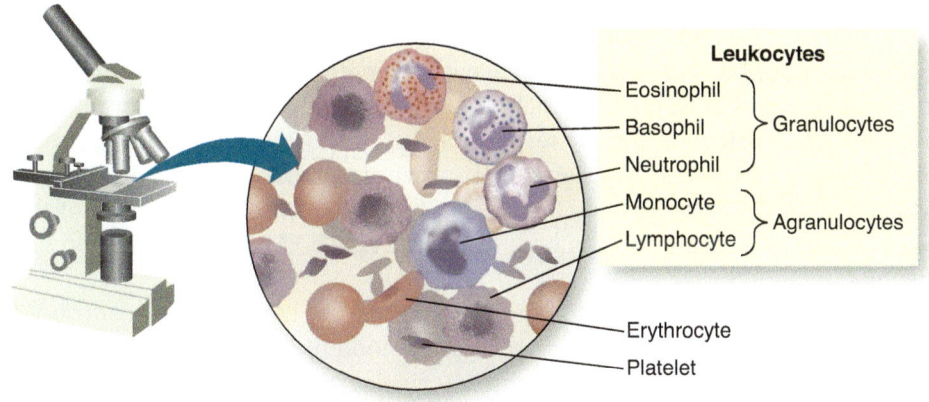

FIGURE 2.3 Several types of blood cells.

referred to as agranulocytes. The **lymphocytes**, of which there are two subsets, the **B lymphocytes** and the **T lymphocytes**, are the key cells of immunity. The B lymphocytes are the only producers of antibodies but are dependent in most cases on **T-helper** (T_H) cells, a subset of T cells, for this function (discussed under Specific Immunity, below).

Neutrophils and **monocytes** are phagocytic cells. Phagocytosis, or "cell eating," is an important defense mechanism by which microbes are engulfed and destroyed; phagocytosis is described in more detail below under Nonspecific Immunity.

A **differential count** reflects the ratio of the white blood cell components and is an important tool in the diagnosis of suspected infection. A drop of blood is smeared onto a microscope slide, covered with a specific stain, rinsed, and examined under the microscope to perform a differential count (Figure 2.3). Frequently,

TABLE 2.2 Categories of White Blood Cells (Leukocytes)

Cell Type	Total Leukocytes (%)	Function(s)
Agranulocytes		
Lymphocytes	25–35	Antibody formation, cell-mediated immunity
B lymphocytes		Produce antibodies
T lymphocytes		
Monocytes	3–8	Phagocytosis
Granulocytes		
Neutrophils	60–70	Phagocytosis
Eosinophils	2–4	Inflammatory response, limited phagocytosis
Basophils	0.5–1	Not clear; contain histamine

in bacterial (versus viral) infections an elevated neutrophil count is found, justifying antibiotic therapy while a more definitive laboratory diagnosis is awaited. By comparison, with the viral disease infectious mononucleosis, there is an increase in lymphocytes.

The role of **basophils** and **eosinophils** is not entirely clear. Eosinophilia (an elevated eosinophil count) is associated with the presence of helminth infections and with allergies. Basophils are rich in granules of histamine that are released during allergic reactions and are responsible for watery eyes, itchy throat, sneezing, and runny nose. The symptomatic treatment is antihistamine medication, of which there are many over-the-counter choices available. You may be one of the unlucky sufferers. Don't feel too bad: Approximately 50% of the population has allergies.

In addition to blood cells, there are blood proteins (Table 2.1) that contribute to immune function. **Complement**, a series of blood proteins, is a significant defense mechanism against potential disease-causing microbes. Complement is a system of **biological amplification.** Consider the role of the amplifier in a music system as an analogy. It does not actually produce the music; if the amplifier were unplugged the music would continue, but the quality would suffer. In much the same way complement enhances phagocytosis, a nonspecific defense mechanism by which bacteria are engulfed and destroyed. Further, complement can bring about lysis (destruction by chemical attack) of the bacterial cell wall, causing it to "spill its guts" through the now-leaky membrane.

Interferon, another component of blood, was discovered in 1957 and was so named because of its ability to interfere with viral replication. It is not a single component but a group of related compounds. The product is released from virus-infected cells and acts to signal other cells of impending danger by triggering them to release virus-blocking enzymes. When interferon was discovered, the medical community had lofty expectation that it would prove to be a powerful antiviral drug that would be as effective against viruses as antibiotics are against bacteria. Unfortunately, interferon has not lived up to expectations. It has been only partially successful in the treatment of certain forms of hepatitis.

Lymphatic System

The **lymphatic system** is anatomically intertwined with the blood circulatory system. **Lymph**, a tissue fluid derived from blood but without blood cells, is circulated throughout the lymphatic capillaries and veins and opens into the blood circulatory system, allowing return of lymphatic fluid. In elephantiasis worms block the lymphatic vessels, leading to the grotesque elephant-like limbs. Hence, the two systems are interdependent: Lymph is derived from blood and drains back into blood (FIGURE 2.4). Situated along the path of the lymphatic veins are the **lymph nodes**. Swollen lymph nodes in the armpits, groin, or neck indicate that these nodes have filtered out invading microorganisms; an infection is in progress. These nodes are highly significant because they contain both B and T cells, enabling antibody production. Lymph nodes also contain phagocytic cells, which destroy microbes and remove damaged or worn-out cells of the body. Thus lymph nodes contribute to both the nonspecific and specific immune systems.

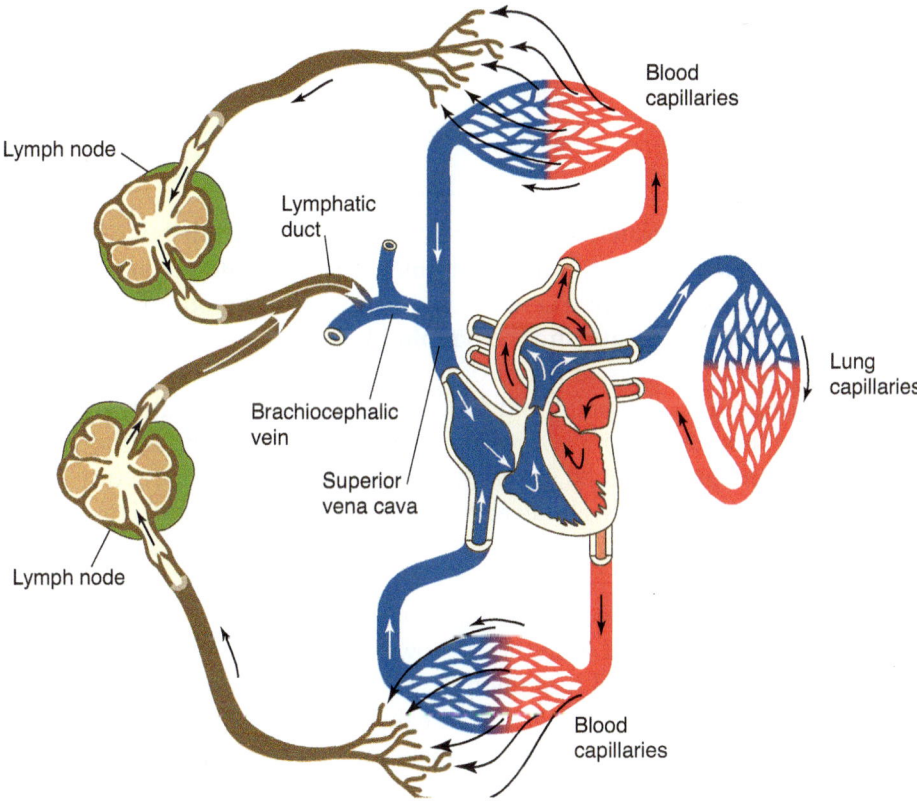

FIGURE 2.4 Interrelationship of the lymphatic and circulatory systems. Lymph is derived from blood and returned to blood via the lymphatic duct.

Primary Immune Structures

Bone Marrow

The **bone marrow**, a **primary immune structure**, plays a highly significant role in immune defense mechanisms; it is the source of all blood cells (red blood cells,

FIGURE 2.5 Pathways of blood cell maturation.

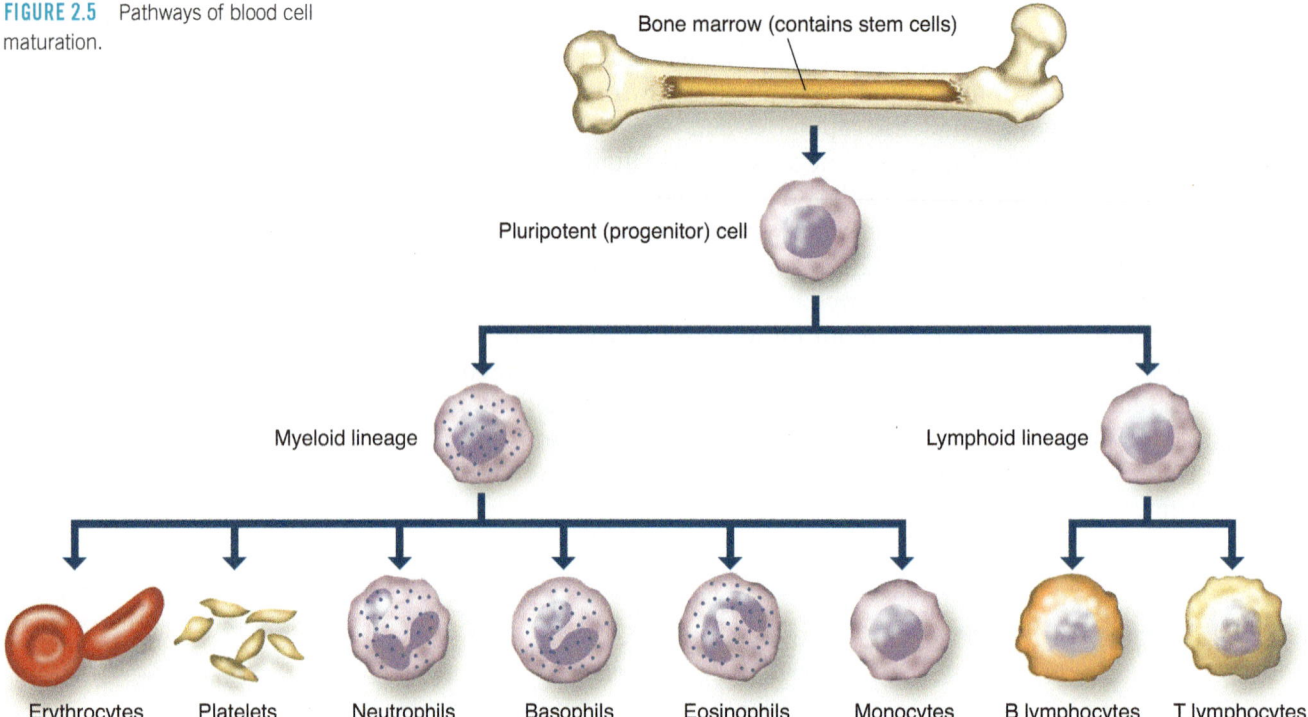

Bone marrow (contains stem cells)

Pluripotent (progenitor) cell

Myeloid lineage

Lymphoid lineage

Erythrocytes   Platelets   Neutrophils   Basophils   Eosinophils   Monocytes   B lymphocytes   T lymphocytes

white blood cells, and platelets) and produces them from stem cells located in the red bone marrow of the sternum (breastbone), vertebrae, and the upper ends of the long bones of the body. There are two pathways of blood cell maturation (FIGURE 2.5), the **myeloid path** and the **lymphoid path**. The myeloid lineage leads to platelets, red blood cells, and four of the five categories of leukocytes (monocytes, neutrophils, eosinophils, and basophils), whereas the lymphoid lineage leads to lymphocytes. B lymphocytes complete their maturation in the marrow and are released into the blood, from which they seed secondary structures of the immune system. Ultimately, all categories of blood cells are released into the general circulation as mature cells.

Thymus Gland

T lymphocytes are released from the marrow before their maturation, and some end up in the **thymus gland** (not thyroid), an organ located behind the sternum and just above the heart (Figure 2.2). In the thymus gland these cells mature into T cells that seed secondary structures of the immune system.

At birth, the thymus is a relatively large organ and attains its maximum size by puberty. The early years are the ones in which numerous childhood diseases and immunizations are experienced; the immune system is maturing, and the thymus plays a significant role. Some children are born without a thymus gland, a condition called **DiGeorge syndrome**. They suffer the consequences of deficient immunity throughout their lifetimes, as manifested by one potentially life-threatening infection after another. After adolescence the thymus gradually degenerates and assumes a lesser significance; its function is taken over by the bone marrow.

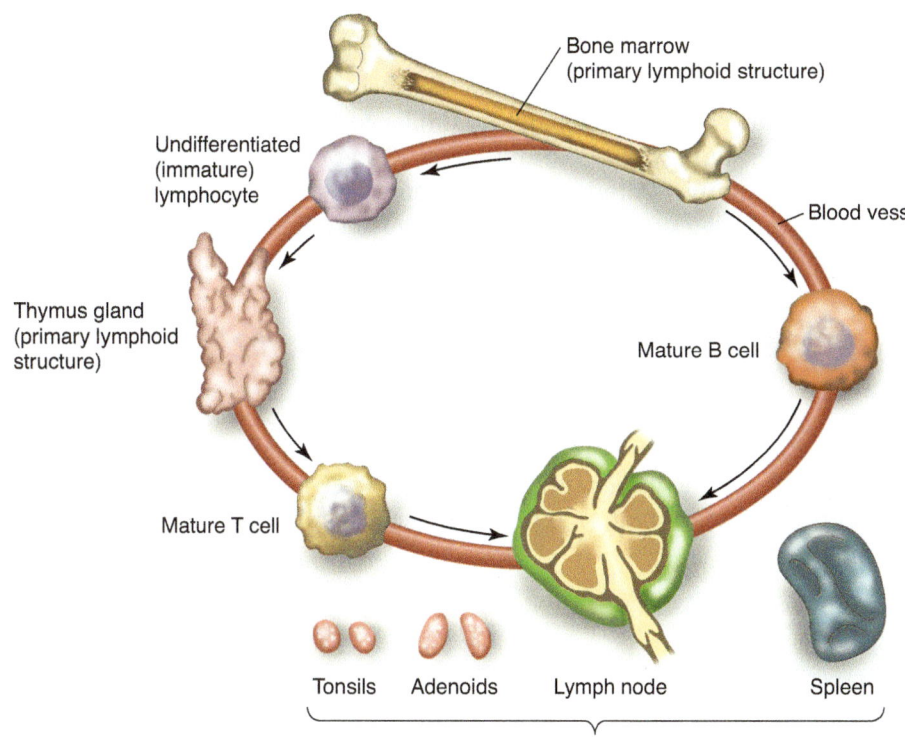

Bone marrow
(primary lymphoid structure)

Undifferentiated
(immature)
lymphocyte

Blood vessel

Thymus gland
(primary lymphoid
structure)

Mature B cell

Mature T cell

Tonsils   Adenoids   Lymph node   Spleen

Secondary lymphoid structures

FIGURE 2.6 Maturation and "seeding" of B and T lymphocytes. B-cell maturation occurs within the bone marrow, whereas other (undifferentiated) lymphocytes mature after migrating from the bone marrow and reach maturity in the thymus gland.

To recapitulate, the thymus is the site of maturation of T cells; the bone marrow is the site of maturation of B cells. These are the primary structures of the immune system and serve as the beds for the maturation of T and B lymphocytes; these mature cells are then seeded into the secondary structures of the immune system (FIGURE 2.6).

Secondary Immune Structures

The **spleen**, **tonsils**, **adenoids**, lymph nodes, and patches of tissue associated with the intestinal tract constitute the **secondary immune structures**. The spleen contains phagocytic cells and both mature T and B cells, endowing it with immunological functions. The spleen is a spongy, fist-sized organ located in the upper left portion of the abdominal cavity. Because of its location and its relatively thin connective tissue covering, it is subject to injury as might be sustained in an automobile accident. Severe trauma to this organ may result in its rupture, with severe hemorrhaging necessitating its removal (splenectomy).

The tonsils and adenoids are located at the back of the throat, just in front of the pharynx. Their lymphocytes play a role in protection against microbes entering through the nose and throat. A generation or so back it was routine for young children to undergo removal of their tonsils (tonsillectomy) because of the frequency of throat and ear infections resulting from infected tonsils; this is no longer a common procedure because of the availability of antibiotics. Doctors in the early 1900s often performed these operations in the home, perhaps on the kitchen table. Occasionally, tonsillectomy is performed on individuals with repeated infections that do not respond well to antibiotics. (As a benefit, you are indulged with a lot of ice cream and popsicles to help alleviate a severe sore throat!) Patches of tissue, called **Peyer's patches**, similar to the tonsils and adenoids, are distributed in the lining of the gastrointestinal, respiratory, and urinary tracts and contribute to immunity. Almost all antigens entering the body (usually as microbial components) end up in lymph or blood and are then carried to lymph nodes or to the spleen, where they encounter cells of the immune system.

■ Duality of Immune Function: Nonspecific and Specific Immunity

Recall that the immune system consists of two components: nonspecific or innate immunity and specific or acquired immunity (FIGURE 2.7). Specific immunity, in turn, has two arms: humoral, or antibody-mediated, immunity and cell-mediated immunity.

FIGURE 2.7 Nonspecific and specific immunity.

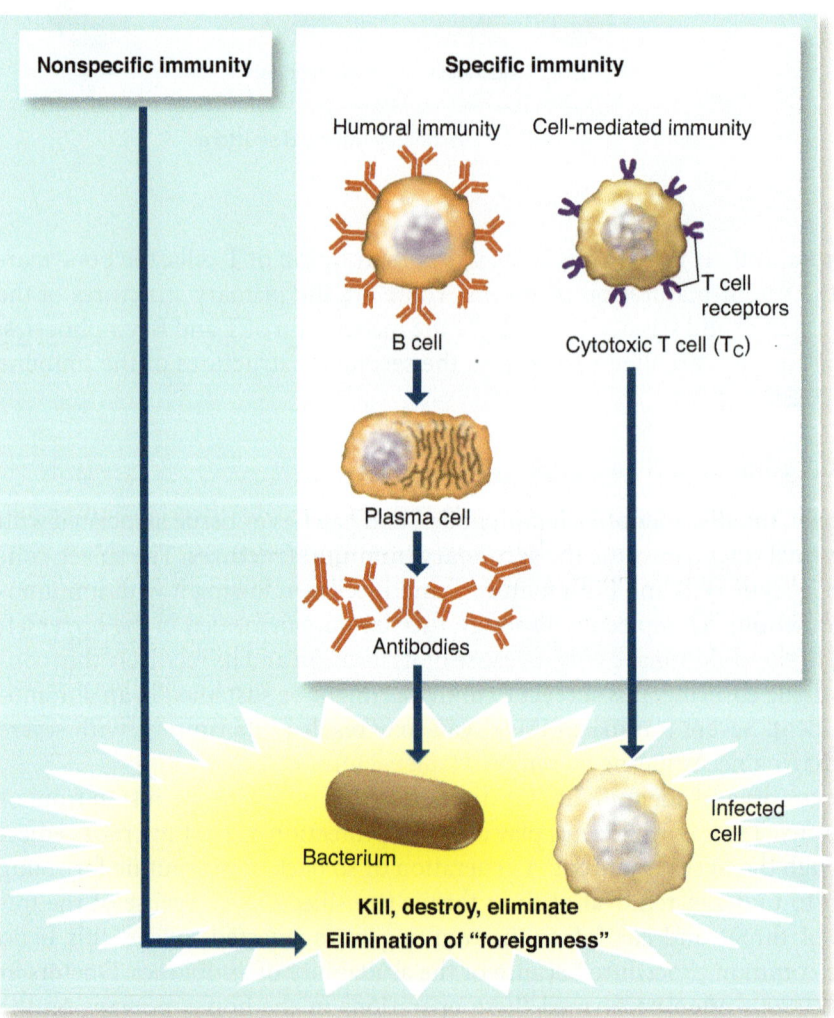

Human Disease and Prevention

Nonspecific Immunity

The foregoing presentation identifies the blood, bone marrow, thymus, and the secondary immune structures as the key elements in the body's protection against foreignness. **Nonspecific immunity** is characterized by physiological defenses that operate either to prevent microbes in the external environment from gaining access into the body or to eliminate those that have penetrated into the body. TABLE 2.3 summarizes nonspecific immunity. These physiological defense mechanisms are present at the time of birth and do not involve specific recognition of the microbe: They act against all microbes in the same fashion. A number of nonspecific internal defense mechanisms, including complement and interferon, as previously described, are antibacterial to those microbes that breach these mechanisms.

TABLE 2.3 Nonspecific Immunity

Mechanism	Function
Skin and mucous membranes	Mechanical barriers
Cilia	Found along respiratory tract and have "upward" motion, pushing microbes up to pharynx where they are swallowed
Phagocytosis	A system of phagocytic cells in the blood and scattered throughout the body
Lysozyme	Found in tears; breaks down bacterial cell walls
Interferon	Found in blood; has antiviral properties
Acid pH of stomach	Many microbes are killed by strong acid environment

Phagocytosis

Phagocytosis is a highly significant nonspecific defense mechanism by which monocytes and neutrophils engulf and destroy foreign substances, including microbes (FIGURE 2.8). Consider the following familiar scenario: Several hours after getting a splinter in your finger, the wound displays the four cardinal signs of **inflammation**: redness (rubor), heat (calor), swelling (edema), and pain (dolor). Microbes from the skin, most likely staphylococci, have invaded through the site of injury. What is the significance of the inflammatory reaction? The injury causes small blood vessels in the area to become engorged with blood (vasodilation), accounting for the redness and the heat; some of the blood leaks from the vessels into the surrounding tissue spaces, causing swelling and pain as a result of increased pressure and the effect of products released by the injured tissue on nerve endings. It hardly seems like a defense mechanism. Within these blood-engorged capillary beds, however, phagocytic cells pile up and stick to the vessel walls surrounding the site of the injury. These cells, attracted in large numbers to the inflamed tissue (chemotaxis), migrate out of the capillaries and kill the bacteria, preventing or minimizing infection. The discovery of phagocytosis by Elie

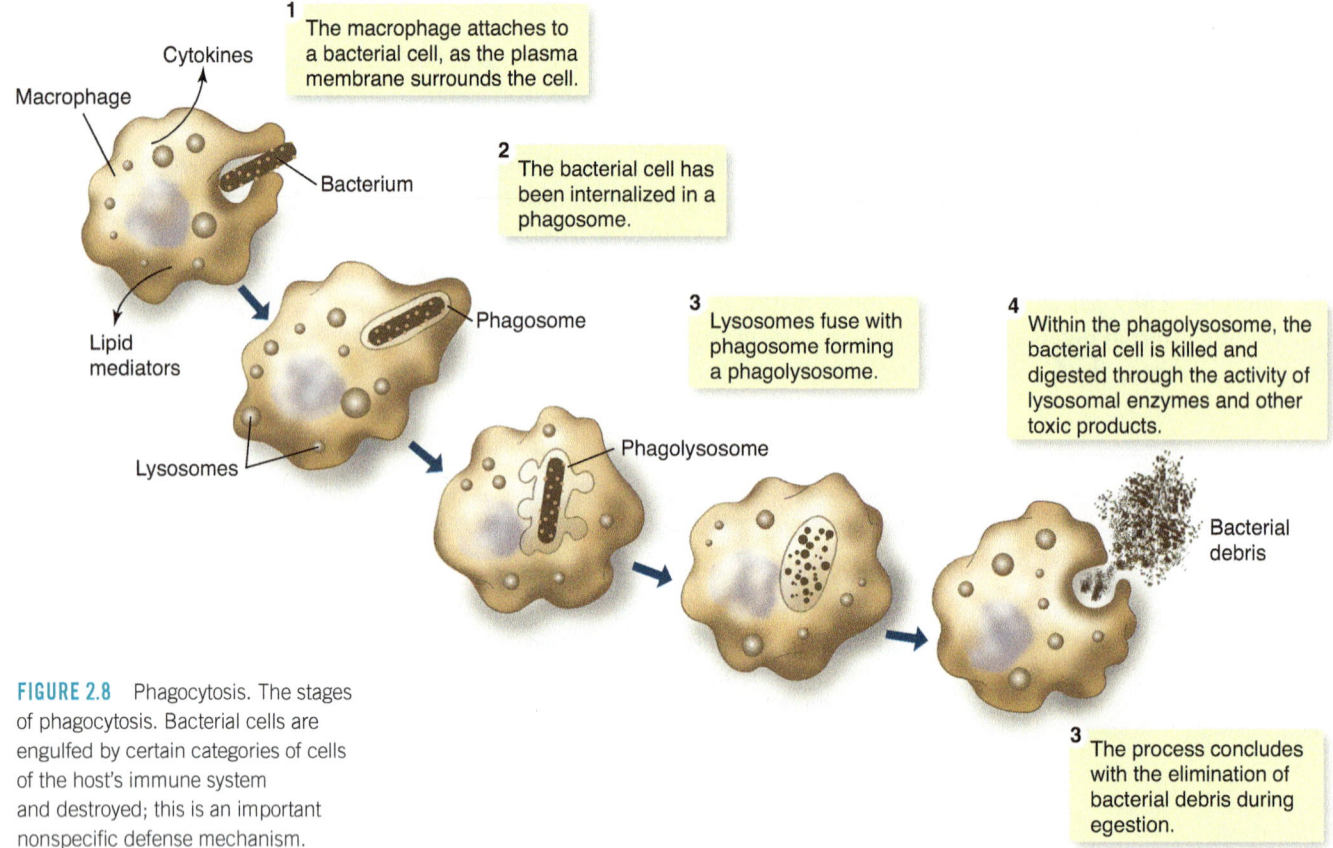

Macrophage

Cytokines

1 The macrophage attaches to a bacterial cell, as the plasma membrane surrounds the cell.

Bacterium

2 The bacterial cell has been internalized in a phagosome.

Lipid mediators

Phagosome

Lysosomes

3 Lysosomes fuse with phagosome forming a phagolysosome.

4 Within the phagolysosome, the bacterial cell is killed and digested through the activity of lysosomal enzymes and other toxic products.

Phagolysosome

Bacterial debris

3 The process concludes with the elimination of bacterial debris during egestion.

FIGURE 2.8 Phagocytosis. The stages of phagocytosis. Bacterial cells are engulfed by certain categories of cells of the host's immune system and destroyed; this is an important nonspecific defense mechanism.

Metchnikoff in 1884 was a landmark in immunology for which Metchnikoff received a Nobel Prize in 1908.

Phagocytosis is an extremely efficient host defense mechanism in its own right but is rendered even more efficient in the presence of antibodies and complement. Phagocytosis is not limited to cells of the bloodstream; a variety of other cells has phagocytic capabilities and is strategically located throughout the body, establishing an efficient system of surveillance. **Macrophages** are monocytes that have migrated out of the blood but retain their phagocytic capability. Some macrophages become fixed at particular sites; other macrophages are called **wandering macrophages** because they move freely about the tissues. Collectively, monocytes, neutrophils, and macrophages are referred to as "professional phagocytes"; they are located strategically throughout the body and play a key role in the destruction of microbes.

Specific Immunity

The specific immune system responds to the presence of foreign microbes that breached the external and internal nonspecific defense mechanisms. At this point an all-out war between microbes and host is in effect, as represented by $D = nV/R$.

A key to understanding immunity is the concept of specificity. Think of antigen-antibody specificity as the complementarity that exists between a lock and a key or between two pieces of a puzzle that fit together.

As previously mentioned, the specific immune system is divided into two categories: **humoral (antibody-mediated) immunity** and **cell-mediated immunity (CMI)**. These two categories are the "big guns." Although these systems operate in very different ways, the mission is a common one: to eliminate foreign antigens—the hallmark of specific immunity. Most antigens are large protein molecules associated with microbes, tumor cells, damaged cells, pollens, dust, and foods. They (1) trigger the production of antibodies specific for that antigen (humoral immunity) or (2) bring about the production of T lymphocytes directed against that antigen (CMI). These lymphocytes are said to be "sensitized."

T Lymphocytes

T lymphocytes have a role in both humoral immunity and CMI, so it important to first understand these cells. There is more than one category of T cells, collectively referred to as the **T-cell subset** (TABLE 2.4). Each category of T lymphocyte has a specific role, identifiable by the **cluster of differentiation (CD)** molecules acquired in the thymus during the T-cell maturation process.

Some T cells, called **cytotoxic T (T_C)** cells, are the effectors of CMI (FIGURE 2.9). The T_C cells are capable of becoming sensitized to ("angry at") the foreign molecules (antigens) carried by the invading microbes. These T cells are identified by **CD8** receptor molecules. The activity of sensitized T cells also depends on **T helper (T_H)** cells.

The T_H cells, identified by **CD4 receptor molecules**, are the regulatory or control cells. The T_H cells are crucial in that the B and T_C lymphocytes are functionally

TABLE 2.4 T-Cell Subsets

Cell Type	Abbreviation	Clusters of Differentiation (Receptors)	Function(s)
T-helper cell	T_H	CD4	Activates B cells to produce antibodies; activates cytotoxic T cells
Cytotoxic T cell	T_C	CD8	Killer cell that works against cells with "foreign" intracellular antigens, including viruses and bacteria
T suppressor cell[a]	T_S	CD8	Regulatory cell works in concert with T_H cells
T delayed-type hypersensitivity cell[a]	T_{DTH}	CD4	Plays role in allergic responses; activates hypersensitivity macrophages

[a]Not discussed in text.

FIGURE 2.9 The protective effects of cell-mediated cytotoxicity. Preforin causes pores to form in the target cell, leading to lysis.

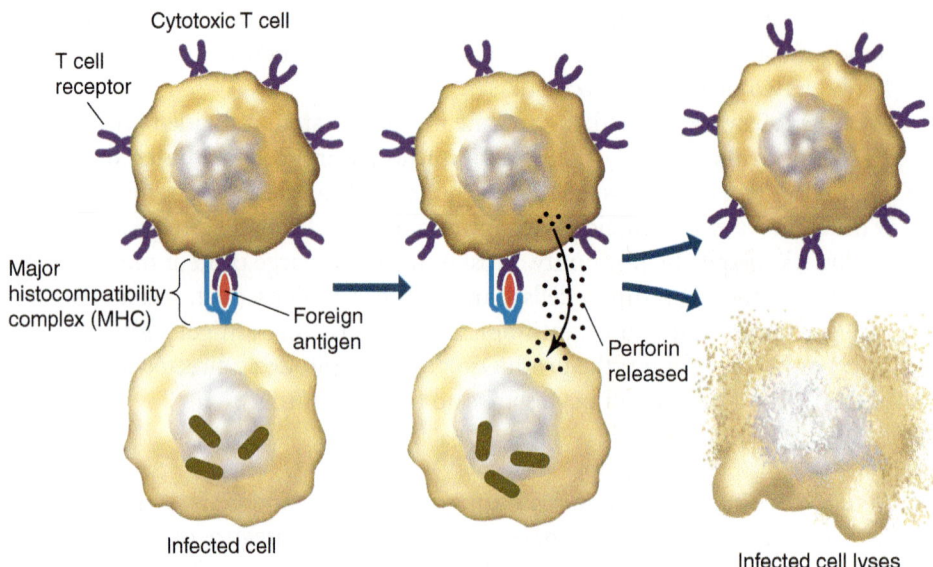

Cytotoxic T cell

T cell receptor

Major histocompatibility complex (MHC)

Foreign antigen

Perforin released

Infected cell

Infected cell lyses

crippled in the reduction or absence of T_H cells. During antigenic stimulation, T_H cells secrete molecules, known as **cytokines**, which activate B and T_C cells.

Antibody-Mediated (Humoral) Immunity

The concept of using an infectious agent to prevent and recover from certain diseases has been realized for centuries. Before the 1700s medical practitioners had some success preventing what later was termed smallpox by picking scabs from pox-infected individuals. The scabs were dried, pulverized, and introduced into the nose of those susceptible. Many of those treated contracted a mild case of smallpox, from which they recovered and were resistant to subsequent infections. In 1796 Edward Jenner introduced a vaccination against smallpox by inoculation of the cowpox virus, resulting in a benign infection in humans. What these early pioneers in immunology did not know was the biological mechanisms behind the resistance—the production of antibodies.

Humoral immunity is mediated by antibodies, products of **B cells** (with the aid of T_H cells) in response to antigens. The antibody molecule is a four-chained structure (**FIGURE 2.10**) that binds with antigens on bacterial cells, viral particles, toxins, and internal cells, including tumor cells and dead cells.

The binding of antibodies to antigens has several outcomes (**FIGURE 2.11**), each of which facilitates destruction of the (antigen-bearing) microbe. B lymphocytes that are "revved up" by contact with antigen and T_H lymphocytes release antibodies and at that stage are called **plasma cells**. Estimates are that a single plasma cell can produce thousands of specific antibodies per second. There are actually five categories of antibodies, or **immunoglobulins (Igs)**, termed **IgG, IgA, IgD, IgE, and IgM**, produced with identical specificity to each antigen. Each type has unique properties. IgG accounts for approximately 80% of the antibody molecules and is the best characterized. The antibody molecule binds with antigens on bacterial cells, viral particles, toxins, and internal cells, including tumor cells and dead cells, resulting in a "kill, destroy, or eliminate" outcome.

28   Human Disease and Prevention

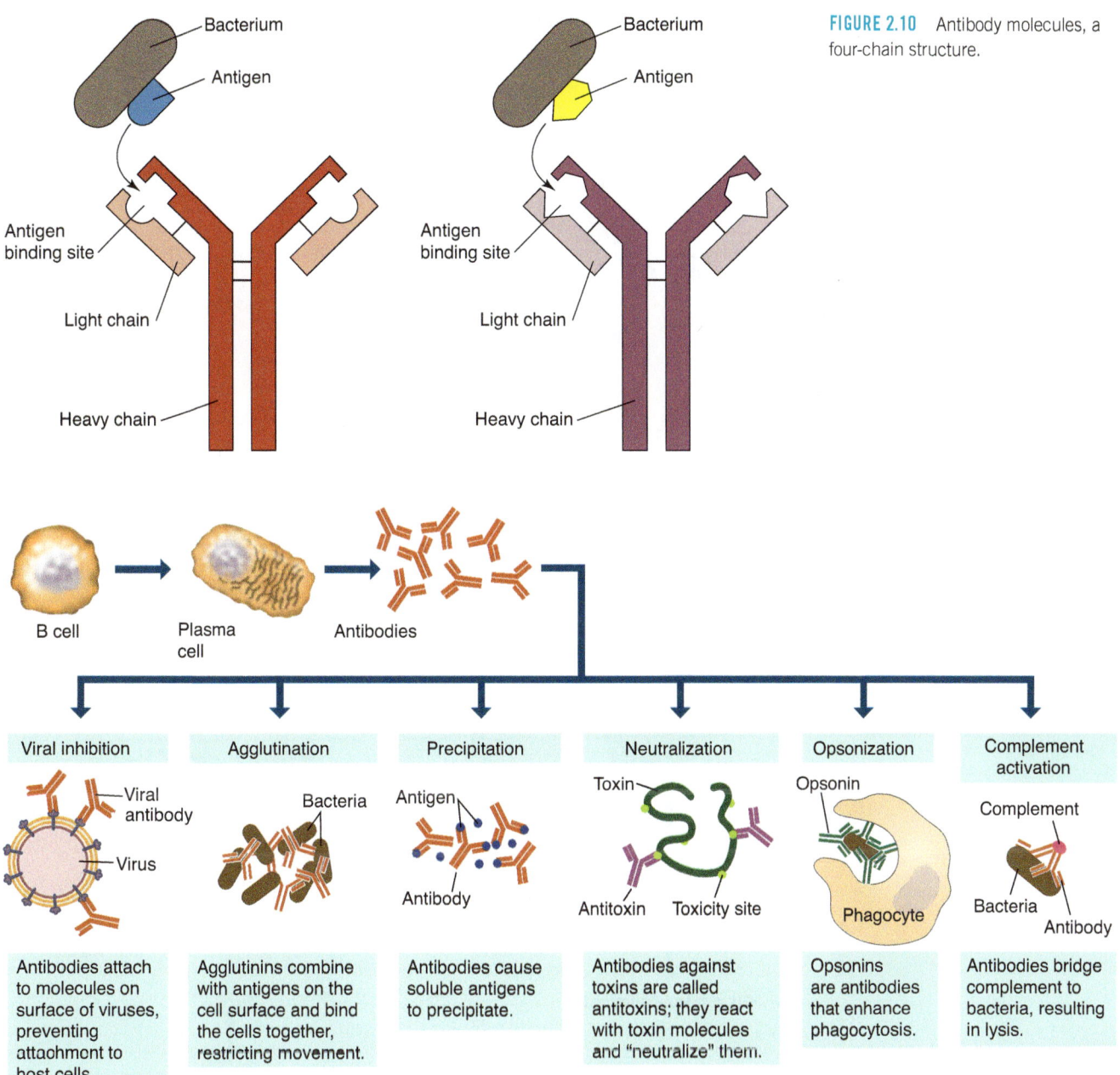

FIGURE 2.10 Antibody molecules, a four-chain structure.

Viral inhibition

Antibodies attach to molecules on surface of viruses, preventing attachment to host cells.

Agglutination

Agglutinins combine with antigens on the cell surface and bind the cells together, restricting movement.

Precipitation

Antibodies cause soluble antigens to precipitate.

Neutralization

Antibodies against toxins are called antitoxins; they react with toxin molecules and "neutralize" them.

Opsonization

Opsonins are antibodies that enhance phagocytosis.

Complement activation

Antibodies bridge complement to bacteria, resulting in lysis.

FIGURE 2.11 Protective effects of antibodies binding to antigens (humoral immunity).

Although it is true that antigen-driven B cells are the only antibody producers, T_H cells are required to help these B cells produce antibodies; they do so by releasing molecules called **cytokines** (FIGURE 2.12). How do B cells know which antibody specificity to produce? The **clonal selection theory** (FIGURE 2.13) is the explanation. According to this theory, a population of B cells for every possible antigen exists at the time of birth. Each B cell displays many copies of a (single) specific antibody molecule on its surface. By random contact antigens "dock" with surface antibodies of corresponding specificity (lock and key), thus stimulating those B cells to

Chapter 2 The Immune Response

29

FIGURE 2.12 Antibody-mediated immunity and CMI.

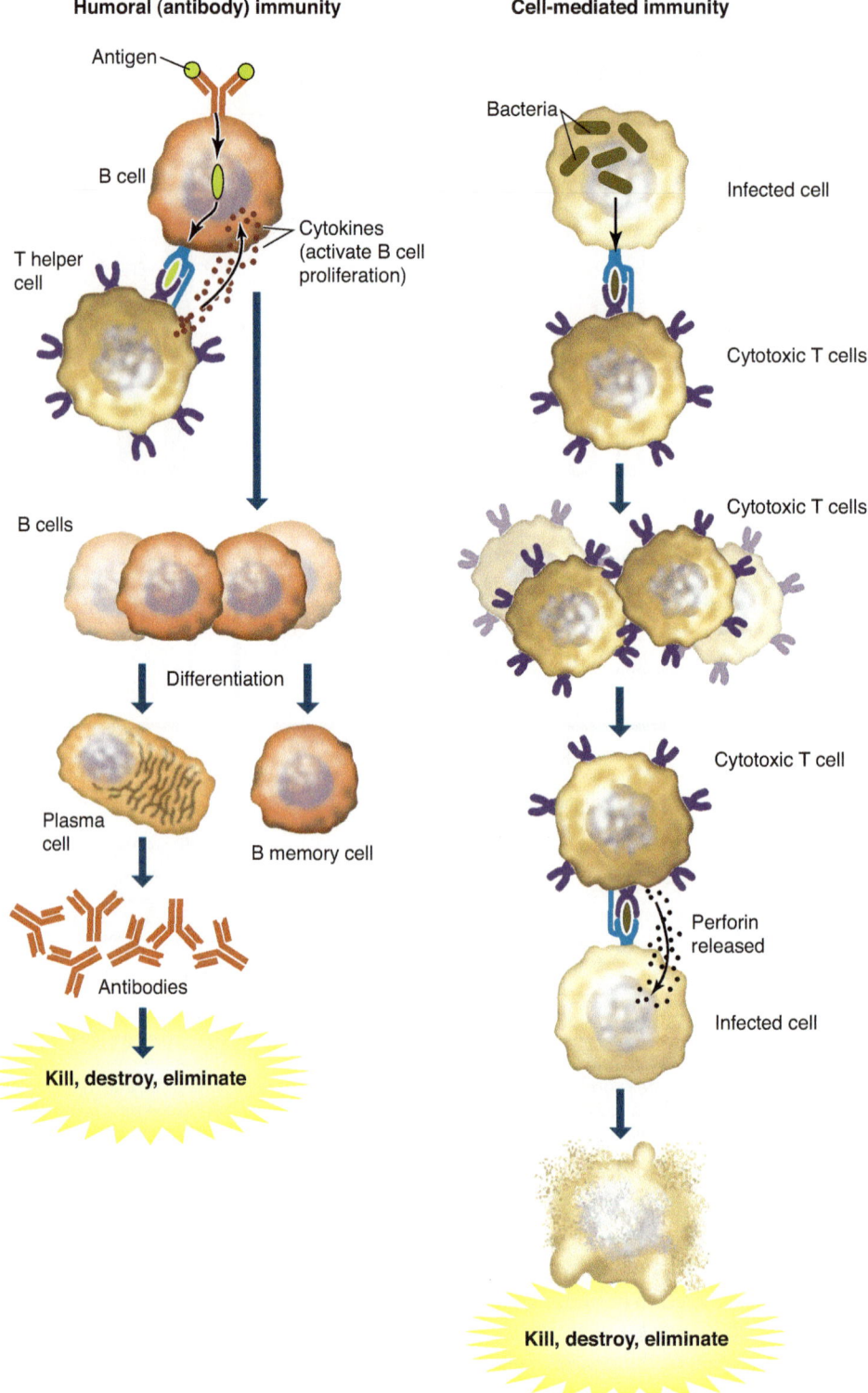

FIGURE 2.13 Clonal selection theory.

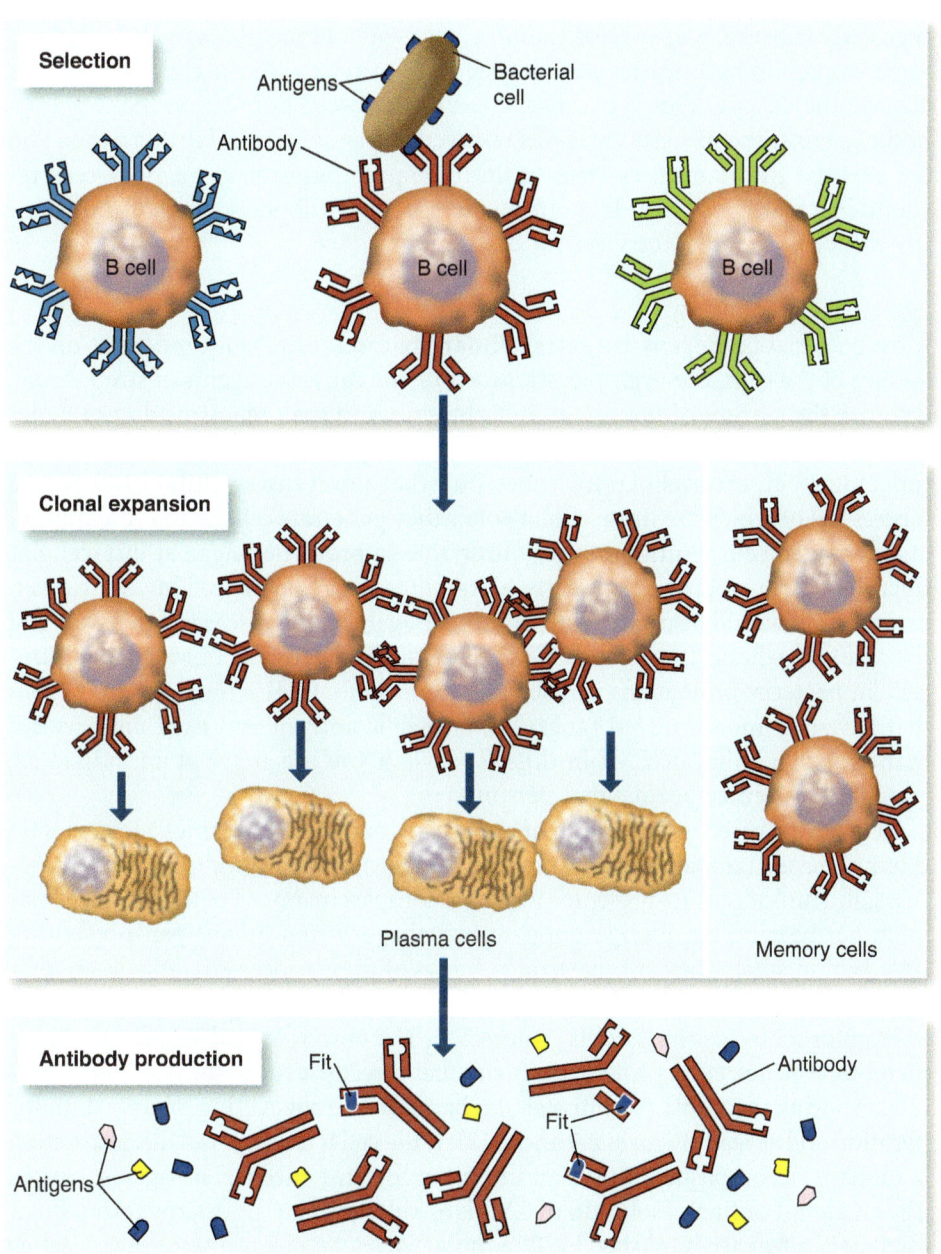

undergo cellular reproduction. Hence, the population of B cells that was selected by the antigen is expanded. Antibodies that are copies of the surface antibody are produced and released. After a lag time of approximately ten to twelve days, antibodies are present in the blood in amounts large enough to be detected.

Some of these B cells may not progress to antibody-producing plasma cells but may remain for years as **memory cells**. This accounts for the fact that, except in rare circumstances, certain so-called childhood diseases (measles, German measles, mumps, and chicken pox) are acquired only once in a lifetime despite repeated exposure. Assume, for example, that you had measles as a child. In later years your child has measles and, despite close contact with the child, you do not acquire the disease. This is because a part of the B-cell population making antibodies against the measles virus was reserved as memory cells. Further, when you are exposed to the measles virus again these preprogrammed memory cells respond in a matter of hours by pouring out measles antibodies at a rate and quantity sufficient to target the virus for destruction.

Cell-Mediated Immunity

Most bacterial pathogens are **extracellular**; that is, they take up residence on the surface of the cells. For example, streptococci, the causative agents of strep throat, colonize the surface of the throat and pharynx, and the organisms do not penetrate into the cells. Antibody-mediated immunity is the body's defense strategy for coping with extracellular microbes, but what about **intracellular** bacteria and viruses (all of which are intracellular) once they penetrate cells or internal mutant and damaged cells? Antibodies play little role in protection against intracellular microbes. Viruses during their early, extracellular stage before cell invasion are an exception (it would be pointless to immunize against viral diseases if this were not the case). So what is the defense? CMI is the major defense strategy against intracellular bacteria, protozoans, viruses, and tumor cells. CMI is mediated by T_C cells that have become sensitized (angry) to a specific antigen and react only against that antigen. (Think of CMI in this way: When you are angry at an individual, your anger is vented against that person.)

Unlike the direct presentation of extracellular antigen to receptor molecules on B lymphocytes (clonal selection theory), antigens from intracellular bacteria, viruses, and tumor cells are presented in a more complex manner. First, the intracellular microbes bearing the foreign antigens are phagocytized by **antigen-presenting cells**, within which they are chewed up. Pieces of these processed antigens are presented on the surface of molecules known as **major histocompatibility molecules** to receptor molecules on T_C cells (Figure 2.9). The now-activated T_C cells produce a membrane-penetrating protein, **perforin**, that directly leads to the lysis of the host cell harboring the viruses or intracellular bacteria, thereby halting microbial multiplication and spread (Figures 2.9 and 2.12). If the cell is a tumor cell, its destruction is similarly accomplished; this is an important defense mechanism against cancer. The end point or final resolution of CMI is to kill, eliminate, or destroy intracellular foreigners, a feat accomplished by T_C lymphocytes.

■ Clinical Correlates

Despite these elaborate mechanisms of defense against potential pathogens, people become ill and sometimes die because of microbial diseases. It may be that the immune system is defeated in the dynamic interplay between microbe and immunity in a virtual tug-of-war by a highly infectious agent such as Ebola or rabies viruses, the bacteria that causes meningococcemia, or HIV. The person's own immune system may go out of control, as is the case with leukemia, or the person may have an immunodeficiency. Treatment with immunosuppressive drugs also increases a person's susceptibility to infection.

Human Immunodeficiency Virus

HIV, responsible for AIDS, destroys T_H cells, resulting in the failure of an infected person to mount an appropriate immune response either by antibody formation or by T_C. The HIV-infected individual becomes severely **immunocompromised** (i.e., has a weakened immune system), which results in one infection after another or several infections at one time, eventually causing death from infection. Blood from AIDS-infected individuals is routinely monitored to determine the level of T_H lymphocytes. The prognosis is grave when the number of T_H cells drops significantly, signaling that the individual no longer has the capacity to combat invading microbes via the specific immune system. When you hear that an individual with AIDS has a low CD4 level, this refers to a low level of T_H cells.

Leukemia

Leukemia, unfortunately, is an all too familiar term. It is a form of cancer characterized by uncontrolled reproduction of white blood cells. There are two major categories of leukemia, reflecting the two pathways of blood cell maturation (Figure 2.5). **Myeloid leukemia** is the result of overproduction of monocytes and granular leukocytes (neutrophils, basophils, and eosinophils). Consider that immature neutrophils may be produced in excess and are unable to carry out phagocytosis, resulting in an immunocompromised individual. **Lymphocytic leukemia** is the result of overproduction of lymphocytes, leading to abnormally large numbers of immature and nonfunctional lymphocytes and their spread by **metastasis** into tissues throughout the body, crowding out normally functioning cells. Leukemia is classified as acute or chronic. In **acute leukemia** the symptoms appear suddenly and progress rapidly; death occurs in a few months unless the condition is successfully treated. **Chronic leukemia** can remain undetected for many months, and life expectancy without treatment is somewhere around three years. Acute lymphocytic leukemia is the most common form of leukemia in children, but fortunately treatment of this condition has a high success rate. The Dana-Farber Cancer Institute (originally the Jimmy Fund Building) in Boston is world renowned for its success in treating children with leukemia. Fans of baseball history will be delighted to know that baseball legend Ted Williams of the Boston Red Sox was a frequent visitor and contributor to this institution.

Immunodeficiencies

A variety of immunodeficiencies may be present at birth as a result of inheritance, including deficiencies in complement production, phagocytosis, B cells, and T cells. A properly functioning immune system requires the interplay of both non-specific and specific mechanisms; impairment results in an immunocompromised individual subject to repeated life-threatening infections. TABLE 2.5 gives a sampling of immunodeficiency disorders, three of which are described here.

Severe combined immunodeficiency (SCID) is a devastating congenital disease in which individuals lack functional T and B cells and therefore cannot mount either an antibody- or cell-mediated immune response. Death is a certainty, resulting from repeated infections. This disorder is commonly known as "bubble

TABLE 2.5	Sampling of Immunodeficiency Disorders

Deficiencies in complement
  C3 (complement factor 3) deficiency
  C5 (complement factor 5) deficiency
Deficiency in phagocytosis
  Chronic granulomatous disease
  Chediak-Higashi syndrome
Deficiencies in B lymphocytes
  X-linked agammaglobulinemia
  Bruton's syndrome
Deficiencies in T lymphocytes
  DiGeorge syndrome
  Wiskott-Aldrich syndrome
  Ataxia telangiectasia
Deficiencies in both B and T lymphocytes
  Severe combined immunodeficiency disorder (SCID)

boy disease," because infants born with this condition must be kept in a germ-free environment, such as a plastic bubble. (One of John Travolta's early movies, *The Boy in the Plastic Bubble* (1976), dramatized the plight of a boy who spent his first twelve years in a germ-free bubble because he was born with SCID.) Those afflicted cannot even have contact with their parents, and all items introduced into the bubble, including air, food, and water, must be sterilized. SCID is the result of a genetically caused enzyme deficiency. In recent years gene therapy and early bone marrow transplants (within three months of birth) have proved to be effective in some cases.

DiGeorge syndrome is a disorder resulting from the absence or incomplete development of the thymus gland. T-cell maturation is abnormal, resulting in impairment of both humoral immunity and CMI.

Chronic granulomatous disease is an inherited disorder of phagocytes. Because of the inability of phagocytes to kill, serious, life-threatening, and persistent infections result.

■ Overview

The immune system is a defense mechanism against invading microbes and, additionally, an internal surveillance system. The ultimate outcome of exposure to pathogenic microbes is the result of the dynamic interplay between the disease-producing (virulence) mechanisms of the parasite and the immune system of the host ($D = nV/R$). The recognition of foreignness is the key element in triggering mechanisms of immune defense. The immune system may inappropriately target self as foreign and mount an immune response against self, resulting in a variety of autoimmune diseases.

Anatomically, the immune system is unique in that it shares cells, tissues, and organs with other functional systems of the body. Blood and blood cells, thymus gland, bone marrow, tonsils, lymph nodes, and spleen are key components of the immune system.

The nonspecific immune system includes the skin, phagocytic cells, components of blood, and ciliated cells that line part of the respiratory tract. Specific immunity responds to the presence of foreign and potentially harmful microbes by antibody-producing (B) cells and by cell-mediated cytotoxic (T_C) cells. Both specific immune mechanisms target microbial invaders for destruction.

Disorders of the immune system may be the result of infection, cancer, or immunosuppressive drugs or may be congenitally inherited. Allergic reactions and autoimmune diseases are adverse reactions of the immune system. Allergic reactions are reactions of hypersensitivity (exaggerated responses). Autoimmune diseases are characterized by the misrecognition of self as nonself (foreign), resulting in the immune system mounting an attack against self.

■ Self-Evaluation

PART I: Choose the single best answer.

1. Which type of white cell is primarily involved in phagocytosis?
   a. neutrophils **b.** eosinophils **c.** basophils **d.** halophils
2. Which one of the following is a primary immune structure?
   a. tonsils **b.** lymph nodes **c.** thymus gland **d.** blood cells
3. SCID disease is characterized by a lack of
   a. B cells **b.** T cells **c.** both B and T cells **d.** phagocytic function
4. Which represents innate, nonspecific immunity?
   a. Ig molecules **b.** acidity of stomach **c.** phagocytosis
   d. more than one of the above
5. The AIDS virus specifically attacks
   a. T_H cells **b.** red blood cells **c.** T_C cells **d.** all of the above
6. Myeloid stem cells give rise to all the following *except*
   a. lymphocytes **b.** neutrophils **c.** platelets **d.** basophils
7. Which of these statements is true regarding the antibody molecule?
   a. It is also called the immunoglobulin molecule. **b.** It is a four-chain molecule **c.** Antibodies "fit" into specific receptors.
   d. Items a, b, and c are all correct.
8. Which one of the following is not considered a secondary immune structure?
   a. spleen **b.** tonsils **c.** bone marrow, **d.** lymph nodes

PART II: Fill in the blank.

1. B cells mature in the _____.
2. Which lymphocyte category is necessary for both CMI and antibody-mediated immunity? _____

3. Name a secondary lymphoid structure. _____

4. Which category of T cells produces perforin? _____

1. What are autoimmune diseases? Discuss the immunological basis of these diseases.

2. Describe the clonal selection theory. What does it explain?

3. An individual with AIDS has a low CD4 count. What does this mean? What is the significance of this low count?

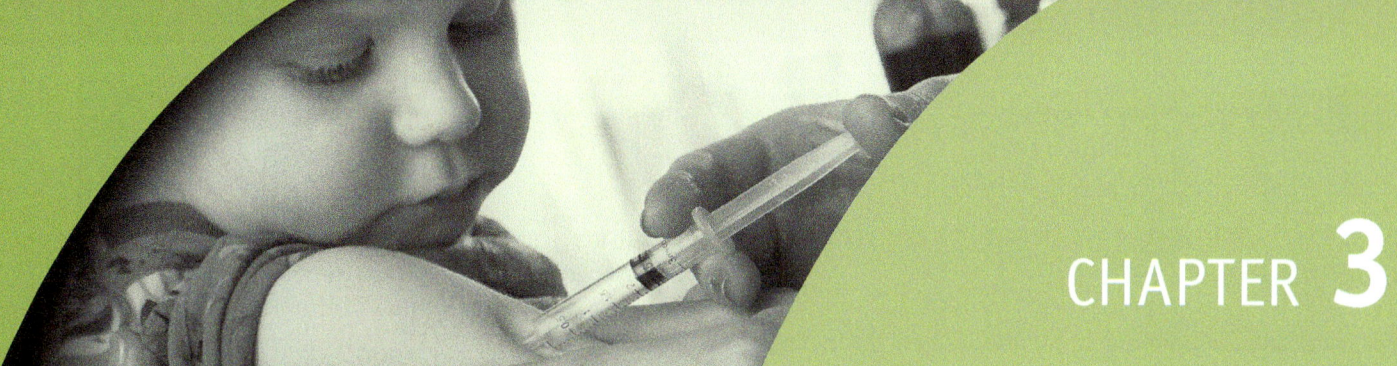

CHAPTER 3

Communicable Diseases

LEARNING OBJECTIVES

By the end of this chapter, the student will be able to:

- describe the burden of disease caused by communicable diseases.
- identify the roles that barrier protections play in preventing communicable diseases.
- identify the roles that vaccinations can play in preventing communicable diseases.
- identify the roles that screening, case finding, and contact treatment can play in preventing communicable diseases.
- identify the conditions that make eradication of a disease feasible.
- describe a range of options for controlling the HIV/AIDS epidemic.

Your college roommate went to bed not feeling well one night and early the next morning you had trouble arousing her. She was rushed to the hospital just in time to be effectively diagnosed and treated for meningococcal meningitis. The health department recommends immediate antibiotic treatment for everyone that was in close contact with your roommate. They set up a process to watch for additional cases to be sure an outbreak is not in progress. Fortunately, no more cases occur. You ask yourself: should your college require that all freshmen have the meningococcal vaccine before they can register for classes?

As a health advisor to a worldwide HIV/AIDS foundation, you are asked to advise on ways to address the HIV and developing tuberculosis (TB) epidemics. You are asked to do some long-range thinking and to come up with a list of potential approaches to control the epidemics, or at least ways reduce the development of TB. The first recommendation you make is to forget about eradicating HIV/AIDS. How did you come to that conclusion?

Your hometown of 100,000 is faced with a crisis as an airplane lands containing a passenger thought to have a new form of severe influenza that has recently gained the ability to spread from person to person through airborne spread. As the mayor of the city, what do you decide to do?

You are a principal at a local high school. One of your top athletes is in the hospital with a spreading bacterial infection due to a staphylococcus bacteria resistant to all known antibiotics. The infection occurred after what appeared to be a minor injury during practice. As the principal, what do you decide to do?

Diseases due to infection form a large part of the history of public health and are again a central part of its present and its future. Infections of public health importance may be caused by a wide variety of organisms ranging from bacteria, to viruses, to a spectrum of parasites including malaria and hookworm. Let us examine the burden of disease due to infections.[a]

---

[a] The term **infectious disease** is intended to include both communicable disease and disease caused by organisms that are not communicable. The central feature that distinguishes **communicable disease** from other diseases caused by organisms is its ability to be transmitted from person to person or from animals or the physical environment to humans by a variety of routes: from air and water, from contaminated articles (fomites), or from insect and animal bites. Other **infections** of public health interest are caused by organisms, such as pneumococcus (streptococcus pneumoniae), which usually coexist with healthy individuals, but are capable of causing disease when relocated to areas of the body with increased susceptibility or when they are provided opportunities to multiply or invade new areas because of an individual's decreased resistance. It is important to note that the distinction between infectious diseases and communicable diseases is not always made and at times they are considered synonyms.

WHAT IS THE BURDEN OF DISEASE CAUSED BY COMMUNICABLE DISEASES?

For many centuries, communicable diseases were the leading cause of death and disability among all ages, but especially among the young and the old. Communicable diseases were not only the causes of great epidemics, but they were the causes of routine deaths. These included a key role in maternal deaths associated with childbirth, infant and early childhood deaths, as well as deaths of malnourished infants and children.

The last half of the 20th century saw a brief respite from deaths and disabilities caused by communicable diseases and other infections. This was due in large part to medical efforts to treat infections with drugs and public health efforts to prevent infections (often with vaccines) and to eradicate or control other infections. Even as these great accomplishments were underway, warning signs of bacteria resistant to antibiotics began to appear. Staphylococcus organisms resistant to current antibi-otics began to plague hospitals in the 1950s until new antibiotics were developed. Resistance of gonorrhea and pneumococcus to a range of antibiotics became widespread. World Health Organization (WHO)—and U.S. government-sponsored pro-grams, such as those to promote the eradication of malaria and TB, were not able to have sustained impacts and the goals were trimmed back to control rather than eradication.

The early 21st century has seen the return of infections that were previously under control, as well as an emergence of new diseases. Tuberculosis, the great epidemic of the 18th and 19th centuries, has returned in force partially as a result of HIV/AIDS. Box 3-1 looks at the history of TB and the histor-ical and current burden of disease caused by it.[1, 2]

Over a dozen previously-unknown infections have emerged in recent decades—the majority of which are believed to have originated in animal species. In the United States, the presence of Lyme Disease and West Nile virus were unknown until the late 20th century, but have now spread to extensive

---

BOX 3-1 The Burden of TB.

"If the importance of a disease for mankind is measured by the number of fatalities it causes, then tuberculosis must be considered much more important than those most feared infectious diseases, plague, cholera and the like. One in seven of all human beings die from tuberculosis. If one only considers the productive middle-age groups, tuberculosis carries away one-third, and often more."[1]

Robert Koch, March 24, 1882

The history of tuberculosis (TB) goes back to ancient times, but beginning in the 18th century it took center stage in much of Europe and America. It has been estimated that in the two centuries from 1700 to 1900, tuberculosis was responsible for the deaths of ap-proximately one billion human beings. The annual death rate from TB when Koch made his discovery was approximately seven million people.[1] Today that would be the equivalent of over 30 million people considering today's population.

Robert Koch's discovery of the association between the tuberculosis bacilli, its culture and isolation, and its transmission to a vari-ety of animal species provided a clear demonstration that the bacilli are a contributory cause of the disease. While the tuberculosis bacilli are clearly a contributory cause of the disease tuberculosis, they are not sufficient alone to produce disease. A large percentage of the world's population harbors tuberculosis. Other factors are needed to produce active disease. These factors include reduced im-munity and nutrition, as well as genetic factors.[b]

The discovery of the tuberculosis bacilli actually followed the development of what was called the sanitarium movement, which began in Europe and America. Sanitariums isolated tuberculosis victims, while providing good nutrition and clean air. The sanitarium move-ment was coupled in the early 20th century with the use of the Bacillus Calmette-Guérin (BCG) vaccine, purified protein derivative (PPD) skin tests, and the recently-invented chest X-ray. These three early and rather crude technologies are still in use today and were de-signed to prevent and diagnose TB. In addition, the understanding of the epidemiology of the tuberculosis bacteria led to a clear vic-tory for public health with the near elimination of TB from the milk supply early in the 20th century.

Thus, even before the ability to actively treat tuberculosis, public health interventions were able to dramatically reduce the frequency of the disease at least in Europe and America. A second round of efforts to control TB began in the 1940s with the discovery of strep-tomycin, the first anti-TB drug, followed over the next decade by para-aminosalicylic acid (PAS) and isoniazid (INH). Combination drug treatments proved highly effective. In addition, INH was found to be effective on its own to prevent skin-test-positive TB from pro-gressing to active disease. Public health and medical efforts to conduct screening for positive skin tests and selectively treating with INH became widespread.

continues

BOX 3-1 continued.

Thus, by the late 1950s and 1960s, TB was brought under control by the combination of medical and public health advances. The success of this effort resulted in the closing of TB sanitariums, gradual cutbacks in screening and treatment programs, and a general loss of interest in TB. From the mid-1960s on, there was little interest or research on TB. Tuberculosis became a treatable disease usually handled as part of routine medical care. Most medical and public health practitioners regarded it as a disease ready for eradication.

Unfortunately, TB was prematurely pronounced dead. It had never come under control in many parts of the world and approximately one-third of those living in developing countries today are estimated to harbor the TB bacillus. That is, they carry TB organisms which may multiply and spread in the future.

Soon after the beginning of the AIDS epidemic that began in the 1980s, active TB came back with a vengeance. AIDS patients with latent TB often developed active TB quite early in their battle with AIDS. They then became contagious to others. Tuberculosis may be more difficult to diagnose and may progress faster in AIDS patients.

Coupled with the return of TB as a public health problem, resistance to TB drugs began to emerge. The problem was successfully combated in the 1990s by the simple public health intervention known as directly-observed therapy or DOT. DOT helped ensure that patients received all the prescribed treatment, thus greatly increasing adherence to effective treatment.

Nonetheless, in recent years resistance to TB drugs and extreme resistance to drug treatments have increased all over the world. Today, we are faced with a triple threat of limited recent research leaving us without modern diagnostic aids or new drugs; a rapidly emerging threat from multiple-drug-resistant tuberculosis, and a spreading epidemic of HIV/AIDS predisposing patients to active tuberculosis. In the past, TB has been exceptionally responsive to a wide range of public health and medical efforts to control its spread. There is hope that with increased awareness and increased further research this will happen again.[2]

---

[b] Establishing that an organism is a contributory cause of the disease traditionally relied on **Koch's postulates.** Koch's postulates hold that in order to definitely establish a cause-and-effect relationship, the following four conditions must be met: 1) the organism must be shown to be present in every case of the disease by isolation of the organism; 2) the organism must not be found in cases of other disease; 3) once isolated, the organism must be capable of replicating the disease in an experimental animal; and 4) the organism must be recoverable from the animal. Ironically, TB cannot be shown to fulfill Koch's postulates. However, a very useful set of **Modern Koch's postulates** has been developed by the National Institute of Allergy and Infectious Disease. It requires evidence of an epidemiological association, isolation, and transmissions to establish that an organism is a contributory cause of the disease.[3]

---

areas of the country. Long-established diseases, such as malaria, are extending their geographic range and further extension is expected with global warming. Influenza is anticipated to return again in pandemic form, as has occurred repeatedly in prior centuries. A pandemic is most likely to occur when ongoing mutations produce new strains capable of person-to-person transmission. Whether you live in a dorm, are a public health professional, a politician, a high school principal, or in almost any profession, the types of questions illustrated in these cases are part of your present and your future.

History suggests that public health and medical interventions have and will continue to have major impacts on the burden of communicable diseases. Let us look at the tools available to address this burden.

WHAT PUBLIC HEALTH TOOLS ARE AVAILABLE TO ADDRESS THE BURDEN OF COMMUNICABLE DISEASES?

A range of public health tools are available to address the burden of communicable diseases. Some of these are useful in addressing non-communicable diseases as well, but they have special applications when directed toward infections. These include:

- Barrier protections, including isolation and quarantine
- Immunizations designed to protect individuals, as well as populations
- Screening and case finding
- Treatment and contact treatment
- Efforts to maximize effectiveness of treatments and prevent resistance

Let us look at each of these tools.

How Can Barriers Against Disease Be Used to Address the Burden of Communicable Diseases?

Examples of barriers to the spread of infections are as old as hand washing and as new as insecticide-impregnated bed nets that have had a major impact on the rate of malaria. Barrier protection, such as condoms, is believed by many to be the most successful intervention to prevent sexually-transmitted diseases. The use of masks may be effective in reducing the spread of disease in healthcare institutions, such as hospitals. The same measures may be preventative in the community at large and are a routine part of winter weather habits in much of Asia.

A special form of barrier protection consists of separating individuals with disease from the healthy population to prevent exposure. As we discussed previously, sanitariums had a major impact in reducing outbreaks of TB in the 19th and first half of the 20th century. Today, once again, we are faced with issues of isolation and occasionally have to legally enforce quarantine.

A second traditional public health approach to infections is the use of immunizations. Let us take a look at a range of ways that immunizations can be used to address the burden of infections.

How Can Immunizations Be Used to Address the Burden of Communicable Disease?

Immunization refers to the strengthening of the immune system to prevent or control disease. Injections of antibodies may be administered to achieve **passive immunity**, which may provide effective short-term protection. **Inactivated (dead)** and **live vaccines (attenuated live)** can often stimulate the body's own antibody production. Live vaccines utilize living organisms that also stimulate cell-mediated immunity and produce long-term protection that more closely resemble the body's own response to infection.

Vaccines are now available for a wide range of bacterial and viral diseases and are being developed and increasingly used to prevent infections as varied as malaria and hookworm.[4] Unfortunately, it has been difficult to produce effective vaccines for some diseases, such as HIV/AIDS. Vaccines, like medications, are rarely 100 percent effective and may produce side effects including allergic reactions that can be life threatening. Live vaccines have the potential to cause injury to a fetus and can themselves produce disease particularly in those with reduced immunity. Some vaccines are not effective for the very young and the elderly. Therefore, the use of vaccines requires extensive investigations to define its effectiveness and safety as well as to identify high risk groups for whom it should be recommended.

For instance college students and military recruits who tend to live in close quarters represent two high-risk groups for meningococcal disease. This bacterial infection can be rapidly life-threatening and when present it requires testing and antibiotic treatment of close contacts. Effective vaccination is now a key tool for controlling this disease. Ideally, vaccination occurs before exposure, however when an outbreak occurs, vaccination of large numbers of potentially exposed individuals living in the surrounding area may be key to effective control. Thus public health uses of vaccines need to consider who should receive the vaccine, when it should be administered, and how it should be administered.

Vaccines administration has traditionally been limited to injections as shots or ingestion as pills. New methods of administration including nasal sprays are now being developed. In addition, it is often possible to combine vaccines increasing the ease of administration. Inactivated vaccines may not produce long term immunity and may require follow-up vaccines or boosters. Thus the use of vaccinations requires the development of a population health strategy which gives careful attention to who, when, and how.

Some infections, especially those viruses that are highly contagious, can be controlled by vaccinating a substantial proportion of the population often in the range of 70 to 90 percent. In this situation, those who are susceptible rarely, if ever, encounter an individual with the disease. This is known as **herd immunity**. When a population has been vaccinated at these types of levels for infections—such as chicken pox, measles, and polio—those who have not been vaccinated are often protected. For some vaccines, such as live polio vaccine, herd immunity is facilitated by the fact that the virus in the vaccine can itself be spread from person to person providing protection for the unvaccinated. Thus, public health authorities are interested in the levels of protection in the community, that is the level of protection of the unvaccinated as well as the vaccinated.

In addition to tools for preventing disease in individuals and populations, public health efforts are often directed at screening for disease and conducting what is called case finding.

How Can Screening and Case Finding Be Used to Address the Burden of Communicable Disease?

Ideally, screening for infections fulfills the same criteria for non-communicable diseases and has played a role in controlling the spread of a number of infections. For example, screening for tuberculosis and syphilis was an effective part of the control of these infections well before they could be cured with antibiotics. Today, screening for sexually-transmitted diseases, including gonorrhea and chlamydia, are a routine part of clinical care. HIV screening has long been recommended for high-risk individuals and for populations with an estimated prevalence above one percent. Today, universal screening is increasingly being adopted as a basic strategy for control of HIV/AIDS.

Screening for communicable diseases has often been linked with the public health practice known as **case finding**. Case finding implies confidential interviewing of those diagnosed with a disease and asking for their recent close physical or sexual contacts. Case finding techniques have been key to the control of syphilis and to a large extent TB both before and after the availability of effective treatment. The advent of effective treatment meant that case finding was of benefit both to those diagnosed with the disease and those located through case finding.

Successful case finding aims to maintain confidentiality. However, when following-up sexual contacts confidentiality is difficult to maintain. The potential for public recognition and the attendant social sigma has inhibited the use of case finding in HIV/AIDS in many parts of the world. The reluctance to utilize case finding may change in coming years as early diagnosis and perhaps early treatment become more effective in controlling the epidemic.

In addition to the use of barrier protections, vaccination of individuals and populations, as well as the use of screening and case finding, public health tools also encompass treatment of those with disease and their contacts.

How Can Treatment of Those Diagnosed and Their Contacts Help to Address the Burden of Communicable Disease?

Treatment of symptomatic disease may in and of itself reduce the risk of transmission. Successful treatment of HIV has been shown to reduce the viral load and thereby reduce the ease of transmission. Similarly, treatment of active TB reduces its infectivity. In addition to direct treatment, a public health tool known as **epidemiological treatment** or treatment of contacts has been effective in controlling a number of communicable diseases. Sexual partners of those with gonorrhea and chlamydia are routinely treated even when their infections cannot be detected. This approach presumably works because early and low-level infections caused by these organisms may be difficult to detect. As we saw in the meningococcal scenario, epidemiological treatment may be the most effective way to halt the rapid spread of a disease.

Contact treatment of HIV/AIDS may become a routine part of controlling the disease. It is already recommended and widely used for treatment of needlestick injuries in healthcare settings. Let us look at one additional public health tool being increasingly used to maximize effectiveness of treatment and prevent resistance.

How Can Public Health Efforts Maximize Effectiveness of Treatment and Prevent Resistance?

In recent years, the impact of antibiotic resistance has become painfully obvious—resistant pneumococcus, gonococcus, and tuberculosis have become widespread. In fact, extremely drug-resistant strains of TB exist in many areas of the world today. Efforts to control resistance have been modest compared to the forces encouraging resistance. Forces encouraging drug resistance include: overuse of prescribed antibiotics, over-the-counter sales of antibiotics in many countries, and widespread use of antibiotics to stimulate modest growth in agricultural animals.

As we have seen, one successful effort known as directly observer therapy (DOT) has led to a more effective treatment of TB. As the name implies, DOT aims to ensure complete adherence to TB treatment by observing individuals taking their daily or at less frequent intervals of treatment. This effort has been credited with success even in areas of drug resistance perhaps on the basis that the body can handle resistant TB if most of the organisms are effectively treated. Efforts are now underway to reduce or eliminate the use of antibiotics for animal growth and to place increased restrictions on the prescribing of highly useful new antibiotics.

The recent emergence of methicillin-resistant staphylococcus aureus (MRSA) outside hospitals has drawn long-overdue attention. It is spreading to communities and beginning to affect otherwise healthy individuals, including athletes and others simply in close physical contact. An effective program to control MRSA will require the use of a range of interventions. This is the situation for many of today's increasingly complex communicable diseases.

Let us now turn our attention to strategies that combine many of the specific public health tools designed to address the problems of infection. We will look at two basic strategies for combating complex infections: elimination and control.

HOW CAN PUBLIC HEALTH STRATEGIES BE USED TO ELIMINATE SPECIFIC COMMUNICABLE DISEASES?

Smallpox was the first human disease to be eradicated. An international effort is hopefully nearing completion to eradicate polio. These two viral diseases are the only ones that have been successfully targeted for eradication. As we have discussed, programs to eradicate TB and malaria have never come close to meeting their goals. Talk of the end of HIV/AIDS is even more unrealistic. Let us see what it takes to successfully eradicate a disease and why so few diseases are on the short list for potential eradication.

The history of smallpox has a unique place in public health. The disease goes back thousands of years and it played a prominent role in colonial America where epidemics often killed a quarter or more of its victims, especially children, and left most others, including George Washington, with severe facial scars for life. The concept of vaccination and the first successful vaccination was developed for smallpox. During the 19th and early 20th centuries, smallpox was largely eliminated from most developed and develo---ping countries through modest improvement on Jenner's basic approach to vaccination, despite the many side effects of this quite crude treatment.[c]

Despite the control of smallpox in most developed countries, there were still over 10 million cases annually of the disease in over 30 countries during the early 1960s. In 1967, the WHO began a campaign to eliminate smallpox. The success of the campaign over the next decade depended on extraordinary organizational management and cooperation, but the prerequisite for success were the unique epidemiological characteristics of smallpox that made it possible. Let us outline the characteristics of smallpox that made eradication possible:[5]

- **No animal reservoir**—Smallpox is an exclusively human disease. That is, there is no reservoir of the disease in animals. It does not affect other species that can then infect additional humans. This also means that if the disease is eliminated from humans, it has nowhere to hide and later reappear in human populations.
- **Short persistence in environment**—The smallpox virus requires human contact and cannot persist for more than a brief time in the environment without a human host. Thus, droplets from sneezing or coughing need to find an immediate victim and are not easily transmitted except by human-to-human contact.
- **Absence of a long-term carrier state**—Once an individual recovers from smallpox, they no longer carry the virus and cannot transmit it to others. Some diseases, like HIV/AIDS and hepatitis B, can maintain long-term carrier states and be infectious to others for years or decades.
- **The disease produces long-term immunity**—Once having recovered from smallpox, very effective immunity is established preventing a second infection.
- **Vaccination also establishes long-term immunity**—As with the disease itself, the live smallpox vaccine pro-

duces very successful long-term immunity. Smallpox has not mutated to become more infectious despite the extensive use of vaccination.

- **Herd immunity protects those who are susceptible**—Long-term immunity from the disease or the vaccine makes it possible to protect large populations. At least 80 percent of the population needs to be vaccinated to interrupt the spread of the infection to the remaining susceptible people.
- **Easily-identified disease**—The classic presentation of smallpox is relatively easy to identify by clinicians with experience observing the disease, as well as by the average person. This makes it possible to quickly diagnose the disease and protect others from being exposed.
- **Effective postexposure vaccination**—The smallpox vaccine is effective even after exposure to smallpox. This enables effective use of what is called **ring vaccination**. Ring vaccination involves identification of a case of smallpox, vaccination of household and close contacts, followed by vaccination of all those within a mile radius of the smallpox case. In the past, households within ten miles were typically searched for additional cases of smallpox. These surveillance and containment efforts were successful even in areas without high levels of vaccination.

The presence of these characteristics makes a disease ideal for eradication. While fulfilling all of them may not be necessary for eradication, the absence of a large number of them makes efforts at eradication less likely to succeed. Table 3-1 outlines these characteristics of smallpox and compares them to polio—the current viral candidate for eradication—as well as to measles. Based upon the content of the table, you should not be surprised to learn that the polio campaign has been much more difficult and has taken much longer than that of smallpox. The potential for a successful measles eradication campaign is still being debated.[d]

Finally, take a look at Table 3-2, which applies these characteristics to HIV infection. It demonstrates why the eradication of HIV/AIDS is not on the horizon.

Unfortunately, eradication of most diseases is not a viable strategy. Thus, public health measures are usually focused on control of infections. In order to understand the range of strategies that are available and useful for controlling communicable diseases, we will take a look at three important and quite different diseases—HIV, influenza A, and rabies.

---

[c] While the vaccine against smallpox is very effective, it has many side effects. The live virus contained in the vaccine can itself cause disease in those vaccinated especially if they have widespread skin disease or have compromised immune system. Today, these side effects might have prevented the widespread use of the smallpox vaccine because it could threaten the lives of the large number of HIV positive individuals many of whom are unaware of their HIV infection. Allergic reactions to the vaccine are also quite common. Allergic reactions to the smallpox vaccine, including inflammation of the lining of the heart, prompted discontinuation of a campaign to vaccinate first responders and healthcare professionals soon after the 9/11 attack.

---

[d] Efforts are underway for the eradication of Guinea Worm, which exhibits a number of favorable characteristics for eradication.

TABLE 3-1 Eradication of Human Diseases—What Makes It Possible?

	Smallpox	Polio	Measles
Disease is limited to humans, i.e., no animal reservoir?	Yes	Yes	Yes
Limited persistence in the environment?	Yes	Yes	Yes
Absence of long-term carrier state?	Yes	Yes—Absent, but may occur in immune-compromised individuals	Yes—Absent, but may occur in immune-compromised individuals
Long-term immunity results from infection?	Yes	Yes—But may not be sustained in immune-compromised individuals	Yes—But may not be sustained in immune-compromised individuals
Vaccination confers long-term immunity?	Yes	Yes—But may not be sustained in immune-compromised individuals Virus used for production of the live vaccine can produce polio-like illness and has potential to revert back to "wild type infection"	Yes—But may not be sustained in immune-compromised individuals
Herd immunity prevents perpetuation of an epidemic?	Yes	Yes	Yes
Easily-diagnosed disease?	Yes—Disease easily identified	Yes/No Disease relatively easy to identify, but large number of asymptomatic infections	No Disease may be confused with other diseases by those unfamiliar with measles
Vaccination effective postexposure?	Yes Postexposure vaccination effective	No Postexposure vaccination not effective	No Postexposure vaccination not effective

WHAT OPTIONS ARE AVAILABLE FOR THE CONTROL OF HIV/AIDS?

HIV/AIDS has been a uniquely difficult epidemic to control. An understanding of the biology of the HIV virus helps us understand many of the reasons for this. The HIV virus attacks the very cells designed to control it. The virus can avoid exposure to treatments by residing inside cells without replicating. Many treatments work by interrupting the process of replication and thus are not effective when replication stops. The virus establishes a chronic carrier state enabling long-term in-fectivity. High mutation rates reduce the effectiveness of drugs, as well as the effectiveness of the body's own immune system to fight the disease.

Despite these monumental challenges, considerable progress has been made by reducing the load of virus through drug treatment and preventing the transmission of the disease through a variety of public health interventions. To appreciate the efforts to control transmission of HIV/AIDS, it is important to understand the large number of ways that it can be transmitted.

TABLE 3-2 Potential for Eradication of HIV/AIDS

	HIV/AIDS
Disease is limited to humans, i.e., no animal reservoir?	No—Animal reservoirs exist
Limited persistence in the environment?	No—May persist on contaminated needles long enough for transmission
Absence of long-term carrier state?	No—Carrier state is routine
Long-term immunity results from infection?	No—Effective long-term immunity does not usually occur
Vaccination confers long-term immunity?	No—None currently available and will be difficult to achieve
Herd immunity prevents perpetuation of an epidemic?	No—Large number of previously-infected individuals increases the risk to the uninfected
Easily-diagnosed disease?	No—Requires testing
Vaccination effective postexposure?	No—None currently available

HIV is most infectious when transmitted directly by blood. Blood transfusions were an early source of the spread of the virus. The introduction of HIV virus testing in the mid 1980s led to dramatic improvement in the safety of the blood supply. Nonetheless, the safest blood transfusions are those that come from an individual's own blood. Thus, donation of one's own blood for later transfusion when needed has become a routine part of surgery preparation in many parts of the world. The most dangerous forms of transfusions are those that come from blood or blood products pooled from large numbers of individuals. Hemophiliacs in many developed countries used pooled blood products to control their bleeding in the 1980s. They suffered perhaps the world's highest rate of HIV infection before this hazard was recognized and addressed. A more recent pooling of blood products occurred in China and contributed to a surge of the disease.

Unprotected anal intercourse is a highly infectious way to transmit HIV. This may help to explain the early spread of the disease among male homosexuals. Today, however, heterosexual transmission is the most common route of infection; additionally, there is a higher risk of transmission from male to female than from female to male. A series of public health interventions have now been shown to be effective: properly used latex condoms, male circumcision, and abstinence are being promoted in efforts to control the disease throughout the world. Aggressive treatment of AIDS at an early stage reduces the viral load and the ease of transmission to others.[e]

Maternal-to-child transmission of HIV was a common, but not universal event before the advent of effective drug treatment. The use of treatment during pregnancy and at the time of delivery has dramatically reduced the maternal-to-child transmission of the infection. Today, this route of transmission is close to being eliminated, which is an important public health achievement. Breastfeeding represents an ongoing and more controversial route of transmission. Up to 25 percent of HIV-positive breastfeeding women may transmit HIV to their children. In countries where breastfeeding provides an essential defense against a wide range of infections, the issue of whether or not to breastfeed has been very controversial. Drug treatment of HIV infections during breastfeeding has been shown to reduce, but not eliminate transmission.

Finally, HIV can be transmitted through contaminated needles. Thus, the risk of HIV transmission needs to be addressed in two very different populations—healthcare workers and those who abuse intravenous drugs. New needle technologies and better disposal methods have reduced the likelihood of needlestick injuries in healthcare settings. Postexposure treatment with drugs has been quite successful in reducing health care-related HIV infections. Reductions in HIV transmission through intravenous drug use have also occurred in areas where public health efforts have focused attention on this method of transmission. Needle exchange programs have met resistance and remain controversial, but most likely contribute to transmission reductions when the programs are carefully designed and administered.

Thus, a range of existing interventions linked to the method of transmission of HIV have been moderately successful in controlling the disease. New methods of control are needed and are being investigated and increasingly applied. Unfortunately, vaccination is not one of them. In fact, randomized clinical trials so far have demonstrated no substantial degree of protection from vaccinations and have raised the concern that vaccination may actually increase the probability of acquiring HIV.

[e] It has been suggested that serial monogamy reduces the risk of HIV transmission in populations compared to having two or more concurrent partners. Serial monogamy contributes to only one chain of transmission at a time and thus may slow, if not halt, the speed or spread of the epidemic in a population.

The recognition that effective vaccinations are not likely in the foreseeable future has brought forth a wide array of ideas on how to control the spread of infection. Antiviral creams, postcoital treatments, and early testing and case finding may become effective interventions. Antiviral creams may become both an adjunct to condom use, as well as a substitute in those situations where condom use is not acceptable. The success of postneedlestick interventions in the healthcare setting has raised the possibility that postexposure treatment may also be effective after high risk sexual contact. Finally, new diagnostic tests for HIV that allow for detection of the disease in the most contagious early weeks of the infection are being investigated for widespread use. To be effective, testing for early disease would need to be coupled with rapid case finding to identify and ideally treat contacts.

It is encouraging to know that existing and emerging interventions for HIV hold out possibility of effective control. Public health and medical intervention complement each other and are both needed if we are to effectively address the most

widespread epidemic of the 21st century. Table 3-3 summarizes the routes of transmission and the estimated transmission rates per exposure.[6] Finally, it outlines the potential interventions that we have discussed.

WHAT OPTIONS ARE AVAILABLE FOR THE CONTROL OF INFLUENZA?

Pandemic influenza is not a new problem. The influenza epidemic of 1918 is estimated to have killed an estimated 50 million people in a world populated with 2 billion people. Today, that would translate to over 150 million deaths. The history of the 1918 influenza pandemic is briefly summarized in Box 3-2.[7] The 1958 pandemic of Asian flu caused a similar, if less deadly pandemic. Thus, we should not be surprised if pandemic flu returns in the coming years.

Influenza A is a viral infection that has long been capable of pandemic or worldwide spread.[f] Its ability to be rapidly

[f] Influenza B can also cause epidemics of influenza, however it is not thought to pose the same hazard of pandemic disease that is possible with Influenza A.

TABLE 3-3 Mode and Chances of Transmission of HIV and Existing Interventions

Route of transmission	Estimated transmission rate per exposure	Potential interventions
Blood transfusion Blood and blood products, such as pooled blood products previously used in U.S. by hemophiliacs	Contaminated blood over 90% chance of transmission; pooling of blood dramatically increases infection	Screening of blood to detect HIV early Use of individual's own blood for surgery
Sexual contact—Anal higher than vaginal, which is much higher than oral	Range from 0.1% to 10% with unprotected receptive anal intercourse posing highest risk Vaginal male to female greater than female to male Circumcision reduces risk by half Other sexually-transmitted diseases may increase risk	Latex condom Circumcision Abstinence Serial monogamy reduces spread compared to two or more concurrent partners
Mother-to-child transmission	15% to 40% higher in developing countries Highest rate of transmission at time of vaginal delivery	Cesarean delivery Drug treatment during pregnancy and at time of delivery for mother and child
Breast-feeding	Very low per exposure, but up to 25% over year or more of breast-feeding	Continuation of drug treatment reduces, but does not eliminate transmission
Needlestick exposures Healthcare occupational risk	Less than 0.5% of HIV positive needlesticks result in transmission	Postexposure treatment with drugs established as effective prevention
Injection drug use	Less than 1% per episode of needle sharing	Needle exchange programs

Source: Data from Population Reference Bureau. Facing the HIV/AIDS Pandemic. *Population Bulletin* 2002: 57(3).

BOX 3-2   The Influenza Pandemic of 1918.

The history of the influenza pandemic of 1918 is summarized by the United States National Archives and Records Administration as follows:[7]

World War I claimed an estimated 16 million lives. The influenza epidemic that swept the world in 1918 killed an estimated 50 million people. One fifth of the world's population was attacked by this deadly virus. Within months, it had killed more people than any other illness in recorded history.

The plague emerged in two phases. In late spring of 1918, the first phase, known as the "three-day fever," appeared without warning. Few deaths were reported. Victims recovered after a few days. When the disease surfaced again that fall, it was far more severe. Scientists, doctors, and health officials could not identify this disease which was striking so fast and so viciously, eluding treatment and defying control. Some victims died within hours of their first symptoms. Others succumbed after a few days; their lungs filled with fluid and they suffocated to death.

The plague did not discriminate. It was rampant in urban and rural areas, from the densely populated East coast to the remotest parts of Alaska. Young adults, usually unaffected by these types of infectious diseases, were among the hardest hit groups along with the elderly and young children. The flu afflicted over 25 percent of the U.S. population. In one year, the average life expectancy in the United States dropped by 12 years.

transmitted through the air from person to person and its short incubation period have made it an ongoing public health problem. It often kills the very young, the very old, and those with chronic illnesses, particularly those with respiratory diseases and suppressed immune systems. In addition, the disease continues to mutate creating new types against which previous infections and previous vaccinations have little or no impact. Thus, new vaccines are required every flu season. Seasonal influenza kills over 30,000 people in the United States alone in the average year despite the increasingly widespread use of vaccinations.

A variety of public health and medical interventions have been and continue to be used to address the current and potential threat posed by influenza. They may well all be needed to address future threats. Let us take a look at these interventions.

Inactivated or dead vaccines have been the mainstay of immunization against influenza. Unfortunately, current technology requires approximately six months lead time to produce large quantities of the vaccine. Thus, influenza experts need to make educated guesses about next year's dominant strains of influenza. In some years, they have been wrong and the deaths and disability from seasonal influenza have increased. New technologies for vaccine production are greatly needed and are being extensively researched.

In recent years, live vaccines administered through nasal spray have been developed and increasingly used. These vaccines are more acceptable than shots to most patients and are now considered safe for a wide range of age groups. They raise the hope of greater acceptance and wider use of influenza vaccinations in coming years.

Medications to treat influenza and modestly shorten the course of the disease have also been developed. Influenza experts view these drugs as most useful to temporarily slow the spread of new strains providing additional time for the development of vaccines to specifically target the new strain. Widespread use of influenza drugs has already resulted in resistance raising concerns that these drugs will not be effective when we need them the most. Efforts are underway to develop new drugs and reserve their use solely for potential pandemic conditions.

Despite our best efforts, influenza is expected to continue its annual seasonal epidemic and to pose a risk of pandemic spread. The use of barrier protection such as masks, isolation methods, and even quarantine has been considered part of a comprehensive effort to control influenza. It is clear that we have a variety of public health methods to help control the impact of the disease. It is likely that we will need all of these efforts and new ones if we are going to control the potential deaths and disabilities due to influenza in coming years.[8] Now, let us look at our last example of the development of public health strategies to control communicable diseases—that of rabies.

WHAT OPTIONS ARE AVAILABLE FOR THE CONTROL OF RABIES?

Rabies is an ancient disease that has plagued human beings for over 4000 years. It is caused by a ribonucleic acid (RNA) virus that is transmitted through saliva of infected animals and slowly replicates. It spreads to nerve cells and gradually invades the central nervous system over a 20- to 60-day incubation period. Once the central nervous system is involved, the disease progresses almost inevitably to death within one to two weeks. Any warm-blooded animal can be infected with rabies, but some species are particularly susceptible—most commonly raccoons, skunks, and bats. Cats and dogs can also be infected and transmit the virus.

A multicomponent vaccination strategy has been very successful in preventing the development of rabies in humans. In most recent years, there have been between one and five fatal cases of rabies per year in the United States despite the persistence and periodic increase in rabies among wildlife populations. Let us take a look at how this quite remarkable control effort has occurred.

The ability to successfully vaccinate humans against rabies after the occurrence of a rabies-prone bite has long been a component of the success of rabies reduction among humans. The use of postexposure vaccination was first demonstrated by Louis Pasteur in 1887 and was used to dramatically save the life of a young victim. Early live vaccines had frequent and severe side effects. They were sequentially replaced by inactivated vaccines grown in animal nerve tissue. These replacement vaccines still led to occasional acute neurological complications and gave the treatment a reputation of being dangerous. The development of a vaccine grown in human cell cultures in the 1970s led to safety records comparable to those of other commonly-used vaccines. Today, over 30,000 rabies vaccination series are administered annually in the United States.

The success of rabies control is a result of a series of coordinated efforts to utilize vaccinations in different settings: Vaccines are administered to individuals who are bitten by suspicious species of wild animals including raccoons, bats, skunks, foxes, and coyotes. Victims of suspected rabies bites by dogs and cats may await the results of quarantine of the animal and observation over a 10-day period. When substantial doubt still exists after this time frame, vaccination is recommended. Laws requiring rabies vaccination of dogs and cats have been enforced in the United States for decades and have greatly reduced the number of reported infections in these animals. Today only ten percent or fewer of suspect rabies-prone bites come from dogs and cats.

Wildlife remains the greatest source of rabies—wildlife epidemics occur with regularity. Rabies-prone bites still occur especially from raccoons, which regularly feed from garbage cans in rural, suburban, and occasionally urban America. Recent development of effective oral vaccinations that can be administered to wildlife through baits have been credited with reducing the number of infected animals, especially those residing in close proximity to humans.

Rabies illustrates the variety of ways that a key intervention—vaccination—can be used to address a disease. As with many complex diseases of public health importance, a carefully designed and coordinated strategy is required to maximize the benefit of available technology. In addition, ongoing research is needed to continue to develop new and improved approaches to the control of communicable diseases.[9]

HIV/AIDS, influenza A, and rabies represent three very different communicable diseases. However, they all require the use of multiple interventions, close collaboration between the public health and healthcare systems, and continuing efforts to find new and more effective methods for their control.

Efforts to control infectious diseases have increased in recent years along with the increase in emerging and reemerging infectious diseases. Technological advances have provided encouragement for the future but at times have raised concerns about the safety of our interventions.[g] New technology, new strategies for applying technology, and new ways to effectively organize our efforts are needed to ensure the effectiveness and safety of our efforts to prevent, eradicate and control communicable diseases.

---

g In recent decades the number of vaccinations used in children has increased dramatically. The potential for new side effect from multiple vaccinations has been of recent concern especially the previous use of a mercury containing compound designed to help preserve vaccines. Concerns have been raised that multiple vaccines have increased the risk of autism. Despite investigations which do not support an association, these concerns have persisted.

Key Words

- Infectious disease
- Communicable disease
- Infections
- Immunization
- Passive immunity
- Inactived (dead) vaccine
- Live vaccines
- Herd immunity
- Case finding
- Epidemiological treatment
- Ring vaccination

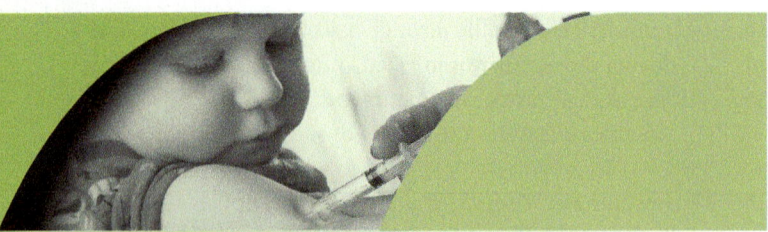

Discussion Questions

Take a look at the questions posed in the following scenarios which were presented at the beginning of this chapter. See now whether you can answer them.

1. *Your college roommate went to bed not feeling well one night and early the next morning you had trouble arousing her. She was rushed to the hospital just in time to be effectively diagnosed and treated for meningococcal meningitis. The health department recommends immediate antibiotic treatment for everyone that was in close contact with your roommate. They set up a process to watch for additional cases to be sure an outbreak is not in progress. Fortunately, no more cases occur. You ask yourself: should your college require that all freshmen have the meningococcal vaccine before they can register for classes?*

2. *As a health advisor to a worldwide HIV/AIDS foundation, you are asked to advise on ways to address the HIV and developing tuberculosis (TB) epidemics. You are asked to do some long-range thinking and to come up with a list of potential approaches to control the epidemics, or at least ways reduce the development of TB. The first recommendation you make is to forget about eradicating HIV/AIDS. How did you come to that conclusion?*

3. *Your hometown of 100,000 is faced with a crisis as an airplane lands containing a passenger thought to have a new form of severe influenza that has recently gained the ability to spread from person to person through airborne spread. As the mayor of the city, what do you decide to do?*

4. *You are a principal at a local high school. One of your top athletes is in the hospital with a spreading bacterial infection due to a staphylococcus bacteria resistant to all known antibiotics. The infection occurred after what*

appeared to be a minor injury during practice. As the principal, what do you decide to do?

REFERENCES

1. Nobelprize.org. Robert Koch and Tuberculosis. Available at: http://nobelprize.org/educational_games/medicine/tuberculosis/readmore.html. Accessed March 18, 2009.

2. New Jersey Medical School Global Tuberculosis Institute. History of TB. Available at: http://www.umdnj.edu/ntbcweb/tbhistory.htm. Accessed March 18, 2009.

3. National Institute of Allergy and Infectious Diseases. The Evidence that HIV causes AIDS. Available at: http://www3.niaid.nih.gov/topics/HIVAIDS/Understanding/HIVcausesAIDS.htm. Accessed March 18, 2009.

4. Centers for Disease Control and Prevention. Vaccines. Available at: http://www.cdc.gov/vaccines/. Accessed March 18, 2009.

5. Sompayrac L. *How Pathogenic Viruses Work.* Sudbury, MA: Jones and Bartlett Publishers; 2002.

6. Lamptey P, Wigley M, Carr D, Collymore Y. Facing the HIV/AIDS pandemic. *Population Bulletin.* 2002; 57(3): 1–3.

7. National Archives and Records Administration. The Deadly Vines: The Influenza Epidemic of 1918. Available at: http://www.archives.gov/exhibits/influenza-epidemic/. Accessed May 12, 2009.

8. World Health Organization. Global Agenda for Influenza Surveillance and Control. Available at: http://www.who.int/csr/disease/influenza/globalagenda/en/index.html. Accessed March 18, 2009.

9. Centers for Disease Control and Prevention. Rabies Home. Available at: http://www.cdc.gov/rabies/. Accessed March 18, 2009.

Epidemiology and Cycle of Microbial Disease

Preview

This chapter considers microbial diseases in relation to public health and focuses on the factors responsible for infectious diseases in populations. Basic concepts of epidemiology are presented so that the occurrence and prevalence of disease in a particular environment at a particular time can be understood. The existence of microbial disease requires a chain of linked factors that constitute the cycle of disease. These factors are reservoir, transmission, portal of entry, and portal of exit.

In hospitals and in long-term health care facilities, all factors involved in the cycle of infectious diseases are present in a concentrated way. These facilities are hotbeds for the transmission of microbes among hospital personnel and patients.

Concepts of Epidemiology

Epidemiology is an investigative methodology designed to determine the source and the cause of diseases and disorders that produce illness, disability, and death in human populations. Epidemiologists have been dubbed "disease detectives"; they are among the first group to be dispatched by the Centers for Disease Control and Prevention (CDC) when the threat of an outbreak occurs anywhere in the world. Their sleuthing is directed at understanding why an outbreak of a disease is triggered at a particular time and in a particular place. Epidemiologists consider age distribution of the population, sex, race, personal habits, geographical location, seasonal changes, modes of transmission, and others. These parameters are used to design public health strategies for control and prevention of future outbreaks. Historically, epidemiology is based on an understanding of the causes and

BOX 4.1   Semmelweis and Childbed Fever

Ignaz Phillip Semmelweiss (FIGURE 4.B1) (1818–1865) graduated from the Vienna University Medical School in 1844 at the age of 26. He started his medical practice in Austria, specializing in obstetrics at a large maternity hospital. He was not an easy doctor to work with. Dr. Semmelweiss was cantankerous, arrogant and abrasive. What he lacked in personality, he made up for in intelligence and his astute observation skills.

The death rate of healthy women from childbed fever was so high that women begged to deliver their babies at home rather than risk a hospital delivery. Semmelweis placed the blame, often with insults, on the doctors' and medical students' unsanitary procedures. He was ridiculed by his colleagues, his hospital privileges were limited, and his academic rank was reduced. "Bad blood" and mysterious forces were thought to be the cause of childbed fever and other diseases; these conditions certainly were not caused by the doctors. Upwards of 30% of women were dying of childbed fever after delivering babies in the hospital whereas women who delivered their babies at home with the aid of a midwife rarely died of childbed fever.

Childbed fever is sometimes called *puerperal fever.* Symptoms of the disease usually began on the second or third day after delivery. The new mothers experienced a violent shivering fit or fever followed by pain in the uterus radiating toward the abdomen that was tender to the touch. The pulse was rapid and the pain became excruciating. Patients complained that the pain was greater than that they suffered during labor. They stopped lactating. As the infection progressed, patients produced cloudy, putrid urine when they could urinate. They produced a foul smelling vaginal discharge. Some vomited and had diarrhea. Their tongues became white, and the patients became thirsty. As the disease progressed, the pain and agony were unbearable. As the suffering continued, they became confused and delirious. Some doctors ordered frequent bleeding or purging procedures and opium for the pain. This was a time before the role of bacteria in diseases was discovered. Today it is known that childbed fever is an infection of the endometrium following childbirth or an abortion caused by group A hemolytic streptococci.

After a woman died of childbed fever, their bodies were moved to nearby dissection rooms—the "death houses." Physicians and medical students would perform autopsies. They smelled of the death houses. They did not wear gloves. Their hands contained putrid matter from corpses. The bloodier and dirtier their laboratory coats, the prouder they became. The smell and filthy coats represented evidence of their superior skills as physicians and interns.

Dr. Semmelweiss observed that the obstetricians and medical students performing autopsies on the wombs of postpartum women who died of childbed fever went directly to the delivery rooms to perform routine vaginal examinations. He believed they carried "death particles" with them to the birthing rooms. Because of his difficult personality, he used his authority to order all obstetricians and medical students to wash their hands with chloride of lime before entering the maternity ward. The results were dramatic! The doctors and interns no longer smelled of death. The morbidity rates plummeted to less than 2%!

Dr. Semmelweiss was ignored and ridiculed by many in the medical community. He believed that cleanliness was critical in the hospitals; however, he could not handle criticism for his beliefs. He was difficult and dogmatic, retaliating by writing angry letters. His term of appointment at the hospital expired in 1849, and he was dismissed. Thirteen years later, he published his treatise, *The Etiology, Concept, and Prophylaxis of Childbed Fever,* which is dated 1861 but was actually published in 1860. The treatise of over 500 pages contains passages of great clarity

distribution of infectious diseases, but modern epidemiology has branched out to other public health problems, including alcohol and drug abuse, cancer, mental conditions, "road rage" and other acts of violence, and exposure to lead paint.

Epidemiology dates back to the time of Hippocrates (460–377 B.C.), who questioned the role of eating and drinking habits, the source of drinking water, lifestyle, and seasons of the year as factors related to causality of disease. He was astute enough to realize that the diseases now identified as yellow fever and ma-

interspersed with lengthy, muddled, repetitive, and belli-cose passages in which he attacks his critics.

In 1865 Semmelweiss was committed to an insane asy-lum. He had become an uncontrollable psychotic—possibly due to tertiary syphilis or Alzheimer's disease. He died at the age of 47, ironically, of a wound infection that may have been caused by the same bacterium that causes childbed fever. A tragic ending for a scientist ahead of his time. Irrespective of his difficult nature, he is credited with the practice of **hand-washing,** a simple, standard aseptic technique. It is the sin-gle most important procedure in preventing hospital-acquired infections. In his honor, Semmelweiss University, a medical school located in Budapest, Hungary, is named after him.

FIGURE 4.B1 Ignaz Semmelweis, sometimes called the "savior of mothers." Before the establishment of microbes as causative agents of disease, Semmelweis realized that childbed fever was transferred from physicians to mothers during delivery. © National Library of Medicine.

AUTHOR'S NOTE
The story of Ignaz Semmelweis is one of my favorites. His life was a struggle to convince his colleagues to wash their hands before delivering babies. I visited the Semmelweis Museum in Budapest during the summer of 2006. There is a small garden in the courtyard of the museum displaying a statue of a mother holding her baby in her arms and raising her eyes in gratitude toward Semmelweis.

Author's photo.

laria were associated with swampy environments where mosquitoes bred. The epidemiological studies of Edward Jenner, a country physician in England, proved that cowpox and smallpox were related and led to smallpox immunization in the late 1700s. In the mid-1800s another early epidemiologist-physician, Ignaz Semmelweis, showed that childbed fever resulted from physicians and other at-tendants who weren't washing their hands after dissecting corpses. They would proceed directly from the autopsy room to the delivery room (BOX 4.1).

FIGURE 4.1 A portion of the map created by Dr. John Snow of the Broad Street area. The lines plot how many people died at each address. Courtesy of Frerichs, R. R. John Snow website: http://www.ph.ucla.edu/epi/snow.html, 2006.

John Snow was another early epidemiologist who laid the groundwork for modern methodologies of epidemiology. In 1849 a major epidemic of cholera occurred in the Soho district of London, causing about 500 deaths in the span of only ten days. Cholera is a bacterial disease manifested by diarrhea so pronounced that life-threatening amounts of water are lost from the body in a short time, causing death from dehydration. Snow's epidemiological detective work showed that most of the cholera victims lived in the Broad Street area and drew their water from the Broad Street pump (**FIGURE 4.1**). Further investigation revealed that the pump was contaminated with raw sewage, and when the pump handle was removed, the cholera epidemic was halted.

A second outbreak of cholera occurred in London in 1854. Snow's sleuthing revealed that most of the cholera victims purchased their drinking water from the Southwark and Vauxhall Company; the company's source of water was the

Thames River downstream from the site where raw sewage was discharged into the river. The Lambeth Company, another water supplier, obtained its water further upstream; the incidence of cholera in the population using Lambeth water was much lower. The cause of the disease was not known and led to conjecture and imagination. The actual bacterial contaminant in the water in both outbreaks proved to be *Vibrio cholerae*.

Epidemiologists focus on the frequency and distribution of diseases in populations and classify diseases as **sporadic**, **endemic**, **epidemic**, and **pandemic**. Sporadic diseases are those that occur only occasionally and at irregular intervals in a random and unpredictable fashion. Typhoid fever, eastern equine encephalitis, and tetanus are examples. Diseases that are continually present at a steady level in a population and pose little threat to the public health are endemic diseases. The common cold, mumps, and whooping cough are endemic across the United States, whereas Lyme disease is endemic primarily in some New England states, but can be caught on a year-round basis. A disease is said to be epidemic when there is a sudden increase in the **morbidity** (illness rate) and in the **mortality** (death rate) above the usual, causing a potential public health problem. Throughout history, epidemics have resulted in more deaths than those caused by wars, and they have influenced the course of history. Plague has bedeviled humankind at least since the reign of Emperor Justinian in the sixth century; the fourteenth century epidemic was particularly devastating. Smallpox, cholera, and typhus fever are other examples of past epidemics. Epidemics may arise from an explosion of sporadic or endemic diseases or, it would seem, from out of nowhere. Pandemic diseases are those that spread across continents and may be worldwide; AIDS is a pandemic. Cholera has been responsible for pandemics on several occasions over time. In 1918 what was then referred to as swine influenza was perhaps the greatest pandemic of all times. Epidemiologists use a variety of graphs, charts, and maps as tools to illustrate the frequency and distribution of diseases (FIGURE 4.2). The term *epidemic* has been borrowed to indicate a variety of conditions unrelated to infectious diseases that are present beyond the norm. For example, college officials talk of grade inflation as a problem of epidemic proportions, as are school violence, obesity, and cancer.

Epidemiologists have described the source and spread of epidemics as **common-source epidemics** or **propagated epidemics**. Common-source epidemics arise from contact with a single contaminated source and are usually associated with fecally contaminated foods and water. Typically, a large number of people become ill quite suddenly, and the disease peaks rapidly in the population. A propagated epidemic is the result of direct person-to-person contact; the microbe is spread from infected individuals to noninfected susceptible individuals. As compared with common-source epidemics, the number of infected individuals rises more slowly and decreases gradually. Chicken pox, measles, and mumps are examples of propagated epidemics. FIGURE 4.3 illustrates the courses of common-source and propagated epidemics.

The number of individuals in a population who are immune (nonsusceptible) to a particular disease as compared with those who are nonimmune (susceptible)

AUTHOR'S NOTE
In 1990 I visited the site of the old Broad Street pump. At the site is now a pub called the John Snow Pub in commemoration of the pump. I had a few beers and reminisced about history.

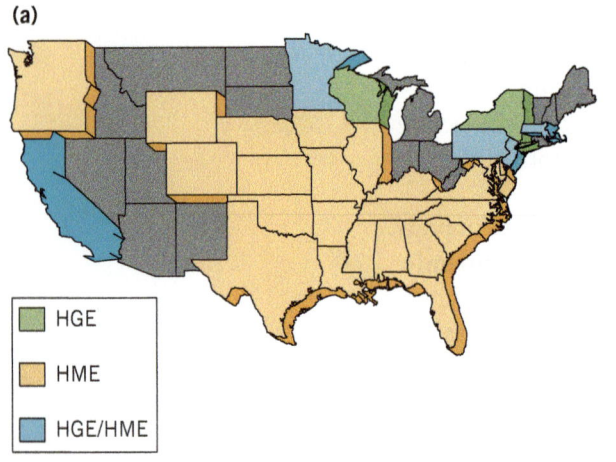

(a)

HGE

HME

HGE/HME

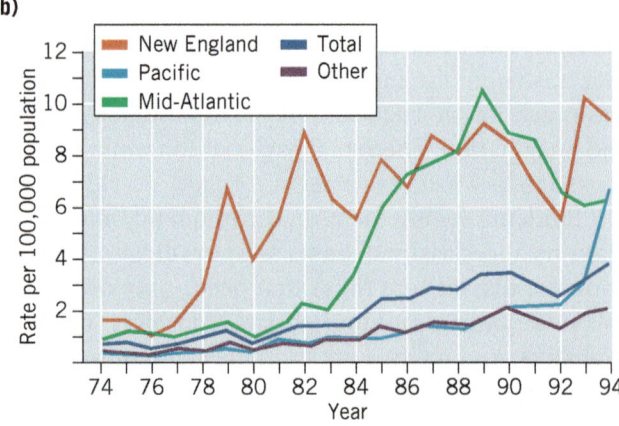

(b)

FIGURE 4.2 Graphs, charts, and maps used by epidemiologists to illustrate disease frequency and distribution. (a) Geographical distribution of human ehrlichioses in the United States. HE, human granulocytic ehrlichiosis; HME, human monocytic ehrlichiosis; gray, no data. (b) *Salmonella enterica* serovar enteritidis isolation rates by region, United States, 1974–1994. (c) Primary causes of chronic liver diseases in a selected area (Jefferson County, Alabama). (Adapted from CDC.)

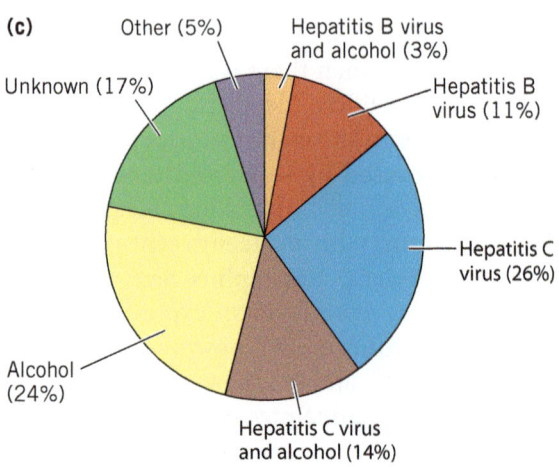

(c)

Other (5%)

Hepatitis B virus and alcohol (3%)

Unknown (17%)

Hepatitis B virus (11%)

Hepatitis C virus (26%)

Alcohol (24%)

Hepatitis C virus and alcohol (14%)

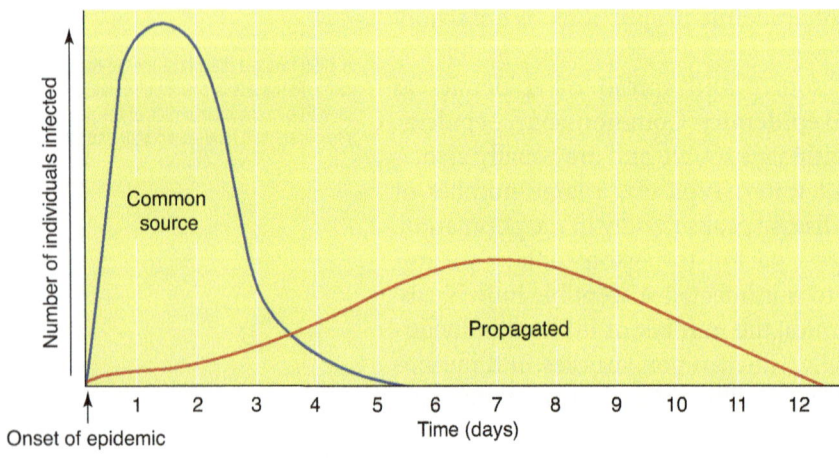

FIGURE 4.3 A comparison of the courses of common-source and propagated epidemics.

is an important factor in the occurrence of epidemics. Immunity can be the result of having had a particular infection or of having been immunized. The term **herd immunity** (group immunity) refers to the proportion of immunized individuals in a population. Disease can only be spread to susceptible individuals; therefore, the smaller the number of susceptible individuals, the less opportunity for contact between them and infected individuals. Public health officials strive to maintain high levels of herd immunity against communicable diseases to minimize the chances that they will progress to epidemic status. Hence, immunization is required in the elementary grades against a variety of diseases; proof of an up-to-date immunization history is required for college admission.

A decrease in herd immunity can lead to reemergence of a disease. A case in point is the epidemic of diphtheria that occurred in the newly independent states of the former Soviet Union in the early 1990s. A decline in the public health infrastructure resulted in fewer children receiving diphtheria vaccination and a decline in herd immunity. When the disease was introduced into the population, possibly by returning military personnel, diphtheria reached epidemic proportions.

Smallpox immunization is no longer practiced because of the eradication of this disease, as proclaimed by the World Health Organization (WHO) in 1980. The world population has little herd immunity against smallpox because only children born before smallpox was declared eradicated in 1980 were vaccinated. This is a potential nightmare, because smallpox virus is near the top of the list of potential biological weapons.

Surveillance of disease outbreaks and of factors that could trigger outbreaks is an important mission of public health organizations throughout the world, including the WHO, the CDC, and agencies at the state and local levels. To keep track of these diseases in the United States, physicians are required to report cases of certain diseases, referred to as notifiable diseases, to their local health departments; these are then reported to the CDC. In 1994 forty-nine diseases were listed as notifiable; sixty-four diseases were reportable in 2008 (TABLE 4.1). The specific diseases are decided on at an annual meeting involving state departments of health and the CDC. An increase or a decrease in the number of notifiable diseases does not necessarily reflect a change in the health status but may be the result of reorganization each year. To further assist public health and medical personnel, the CDC publishes the journal *Emerging Infectious Diseases* as well as the *Morbidity and Mortality Weekly Report,* which contains data organized by states on morbidity and mortality of particular diseases in the United States and throughout the world.

Cycle of Microbial Disease

For infectious diseases to exist at the community level, a chain of linked factors needs to be present, somewhat reminiscent of a parade of circus elephants linked trunk to tail. These factors are reservoirs, modes of transmission, portals of entry, and portals of exit (FIGURE 4.4). An understanding of these factors is imperative to attempt to break the cycle somewhere along the path. For example, if insects are involved in transmission, then controlling their population is a target; for those microbes transmitted by drinking water, providing safe drinking water is a goal. Shrinking the reservoir (where the microbes exist in nature) is a potential target for other diseases. In some instances, a combination of targets is preferable.

For a particular microbial disease to exist there has to be a pathogen as the causative agent and a host in which the pathogen takes up residence. The potential for disease to occur and its outcome are a result of the complex interaction between the

TABLE 4.1 Major Infectious Diseases Designated as Notifiable at the National Level, United States, 2008

AIDS

Anthrax

Arboviral neuroinvasive and nonneuroinvasive diseases

Botulism

Brucellosis

Chancroid

Chlamydia trachomatis, genital infections

Cholera

Coccidioidomycosis

Cryptosporidiosis

Cyclosporiasis

Diphtheria

Ehrlichiosis/anaplasmosis

Giardiasis

Gonorrhea

Haemophilus influenzae, invasive disease

Hansen's disease (Leprosy)

Hantavirus pulmonary syndrome

Hemolytic uremic syndrome, postdiarrheal

Hepatitis viral, acute

Hepatitis virus, chronic

HIV infection

Influenza-associated pediatric mortality

Legionellosis

Listeriosis

Lyme disease

Malaria

Measles

Meningococcal disease

Mumps

Novel influenza A virus infections

Pertussis

Plague

Poliovirus infection, nonparalytic

Poliomyelitis, paralytic

Psittacosis

Q fever

Rabies

Rocky Mountain spotted fever

Rubella

Rubella, congenital syndrome

Salmonellosis

Severe acute respiratory syndrome-associated Coronavirus (SARS-CoV disease)

Shiga toxin-producing *Escherichia coli* (STEC)

Shigellosis

Smallpox

Streptococcal disease, invasive, group A

Streptococcal toxic shock syndrome

Streptococcus pneumoniae, drug-resistant invasive disease

Streptococcus pneumoniae, invasive disease nondrug resistant, in children less than five years of age

Syphilis

Syphilis, congenital

Tetanus

Toxic shock syndrome (other than streptococcal)

Trichinellosis (trichinosis)

Tuberculosis

Tularemia

Typhoid fever

Vancomycin-intermediate *Staphylococcus aureus* (VISA)

Vancomycin-resistant *Staphylococcus aureus* (VRSA)

Varicella (deaths only)

Varicella (morbidity)

Vibriosis

Yellow fever

Reproduced from the CDC. *Summary of Notifiable Diseases,* 2008.

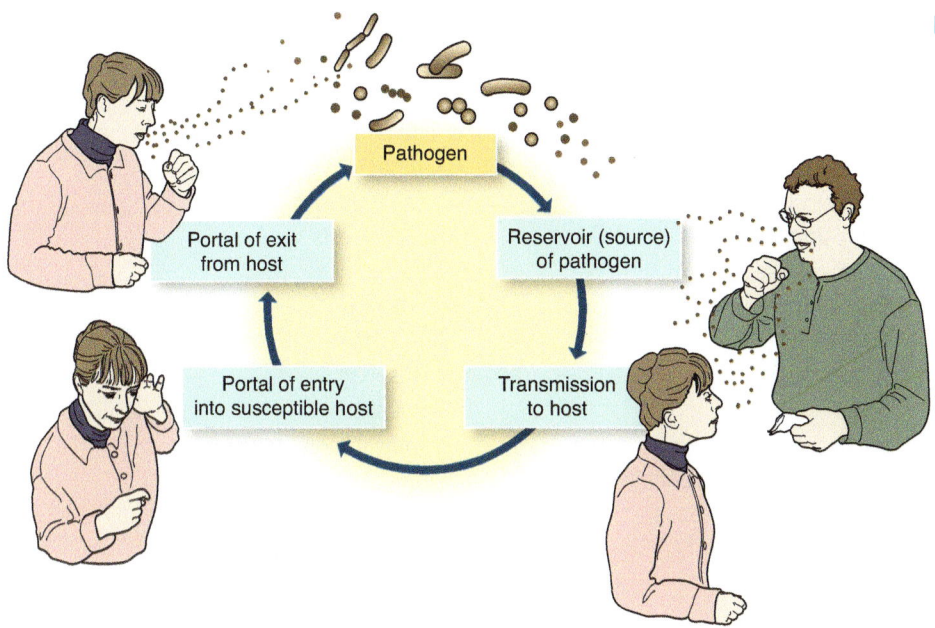

FIGURE 4.4 The cycle of microbial disease.

Pathogen

Portal of exit from host

Reservoir (source) of pathogen

Portal of entry into susceptible host

Transmission to host

number of invading microbes and their virulence and the host immune system. Communicable diseases are infectious diseases in which the pathogen can be transmitted from its reservoir to the host portal of entry.

Reservoirs of Infection

A **reservoir** is a site in nature in which microbes survive (and possibly multiply) and from which they may be transmitted. All pathogens have one or more reservoirs, without which they could not exist. Knowledge and identification of these reservoirs are important, because the reservoirs are prime targets for preventing, minimizing, and eliminating existing and potential epidemics. The facts that humans are the only reservoir of smallpox and that person-to-person transmission of smallpox takes place were key factors in the eradication of this disease. Additionally, humans are the only known reservoir for gonorrhea, measles, and polio. Animals, as well as plants and nonliving environments, also serve as reservoirs. In some cases the source of the pathogen is distinct from the reservoir and is the immediate location from which the pathogen is transmitted. For example, in typhoid fever the reservoir may be an individual with an active case of the disease who sheds typhoid bacilli in feces; the immediate source would be water or food contaminated with fecal material. On the other hand, in most sexually transmitted diseases the human body serves as both reservoir and source.

Active carriers are those individuals who have a microbial disease, whereas **healthy carriers** have no symptoms and unwittingly pass the disease on to others. **Typhoid Mary**, a cook and healthy lifetime carrier of typhoid fever, was responsible for about ten outbreaks, fifty-three cases, and three deaths due to typhoid fever during her lifetime. **Chronic carriers** are those who harbor a pathogen for long periods after recovery, possibly throughout their lives, without ever again

TABLE 4.2 Selected Zoonotic Diseases

Transmission by arthropod bites	Transmission via food, water, or animal bites
Bacteria	Bacteria
Ehrlichiosis	Undulant fever
Relapsing fever	Leptospirosis
Lyme disease	Anthrax
Rocky Mountain spotted fever	Cat scratch fever
Plague	Tularemia
Typhus fever	Viruses
Viruses	Rabies
Yellow fever	Hantavirus disease
Eastern equine encephalitis	Viral gastroenteritis
West Nile virus disease	Protozoans
Rift Valley fever	Giardiasis
Dengue fever	*Cyclospora* infection
Protozoans	Toxoplasmosis
Babesiosis	
Sleeping sickness	
Malaria	
American trypanosomiasis	
Leishmaniasis	

becoming ill with the disease. In the case of chronic (and healthy) carriers of typhoid fever, removal of the gallbladder may be effective in eliminating the carrier state; intensive therapy with antibiotics works in other cases. Tuberculosis is another disease in which carriers play a significant role. Depending on the particular infection, carriers discharge microbes via portals of exit, including respiratory secretions, feces, urine, and vaginal and penile discharges.

Domestic and wild animals serve as reservoirs for about 150 species of pathogenic microbes that can affect humans. These diseases are referred to as **zoonoses** (TABLE 4.2). Microbes of animals that are most closely related to humans have the greatest chance of making the "species leap" to humans or as having erased the species barrier. Consider, for example, the AIDS virus, which is thought to have a reservoir in chimpanzees and is now a human pathogen. Prions cause both mad cow disease and its human counterpart variant, Creutzfeldt-Jakob disease; prions jumped from cattle to humans. Monkeys are reservoirs for the microbes that cause malaria, yellow fever, and numerous other diseases. The reservoirs for the bacteria that cause Lyme disease, a major problem in the northeastern United States, are deer and mice. Hantavirus pulmonary syndrome, a relatively new disease in the United States, uses a variety of rodent species, particularly the deer mouse, as reservoirs. Many mammals, including dogs, cats, raccoons, skunks, foxes, and bats, serve as reservoirs for rabies.

Eradication of zoonotic diseases is particularly challenging and difficult because it is, ultimately, dependent on eradicating the reservoirs. Malaria is a protozoal disease transmitted by mosquitoes. Intensive spraying with the pesticide DDT (dichlorodiphenyltrichloroethane) in the 1940s markedly reduced the mosquito population and the number of malaria cases. The mosquitoes developed resistance to DDT eventually, leading to a reemergence of malaria.

Nonliving Reservoirs

Some organisms are able to survive and multiply in nonliving environments. Soil and water are the major nonliving reservoirs of infectious diseases. The tetanus bacillus and the botulinum bacillus, both members of the same bacterial group *Clostridium,* are spore formers and thus can survive for many years in soil. These organisms are part of the normal flora of horses and cattle and are deposited in their feces onto the soil. The use of animal fertilizers contributes to their distribution. Certain helminth (worm) parasites (for example, hookworms) deposit their eggs onto the soil, establishing a reservoir for human infection (FIGURE 4.5).

Contaminated drinking water and foods are major reservoirs for many microbes that cause gastrointestinal tract disease, ranging from mild to severe to fatal. The list includes bacteria, viruses, and protozoa (TABLE 4.3). Because of the potential for an outbreak of waterborne and foodborne illnesses, local departments of public health devote considerable attention to sanitary measures designed to minimize risks; their activities include monitoring food service establishments, beaches, swimming pools, and certification of food handlers.

FIGURE 4.5 Soil can be a reservoir for microbes and helminth eggs. This child is at increased risk for infection. © Bernardo Erti/Dreamstime.com.

Transmission

The next link in the cycle of disease is **transmission**, the bridge between reservoir and portal of entry (Figure 4.4). Transmission is the mechanism by which an infectious agent is spread through the environment to another person. More simply put, transmission answers the question "How do you get the disease?" There are several modes of transmission, and they can be grouped into two major pathways, direct and indirect. Each of these, in turn, can be subgrouped into three categories (TABLE 4.4).

TABLE 4.3 Water and Food as Reservoirs of Infection

Type of Microbe	Examples of Water- and Foodborne Infections
Bacteria	Salmonellosis, cholera, *E. coli*
Viruses	Hepatitis A, poliomyelitis, viral gastroenteritis
Protozoa	Giardiasis, amebiasis, cryptosporidiosis
Worms	Ascariasis, trichinellosis, *Trichuris* infection

TABLE 4.4 Modes of Transmission	
Direct	**Indirect**
Contact (e.g., kissing, sneezing, coughing, singing, sexual contact)	Vehicles (fomites, e.g., doorknobs, eating utensils, toys)
Animal bites	Airborne (via aerosols created by, e.g., shaking bedsheets, sweeping, mopping)
Transplacental	Vectors (e.g., mosquitoes, ticks, flies)

Direct Transmission

The most common type of **direct transmission** is person-to-person contact, in which the infectious agent is directly and immediately transferred from a portal of exit to a portal of entry. Sexual contact, kissing, and touching are the most common examples. Transmission is facilitated by contact of the warm, moist mucous membranes of one individual with the warm, moist mucous membranes of another, as occurs in sexually transmitted diseases. Sexual contact is an example of **horizontal transmission** (i.e., transmission from one person to another). Droplet transmission is also direct and horizontal and involves the projection of infected spray from coughing, sneezing, talking, and laughing onto the conjunctivae of the eyes or onto the mucous membranes of the nose or mouth. Influenza, whooping cough, and measles are spread by droplets. Droplets are about 10 micrometers or greater in diameter and travel less than 1 meter; as many as twenty thousand droplets may be produced in a sneeze (FIGURE 4.6). Think about that the next time you cough or sneeze directly into the crowded environment of the classroom and be certain to "spray" into the crook of your elbow and not into your hand.

A second type of direct and horizontal transmission involves animal bites, rabies being the most common example. The virus is directly transmitted from the saliva of the rabid animal onto the skin and underlying tissues. Finally, transplacental transmission is an example of **vertical transmission**, in which the pathogens are passed from mother to offspring across the placenta (AIDS, measles, and chicken pox), in breast milk, or in the birth canal (syphilis and gonorrhea). Notice there are no intermediaries in all these categories of direct transmission. Microbes are transferred by the contact of portals of exit with portals of entry. These portals are described later in this chapter.

FIGURE 4.6 Droplet transmission. As many as twenty thousand droplets may be produced during a sneeze. It is important to carry a handkerchief or tissue and to cover your nose and mouth when sneezing. © James Klotz/ ShutterStock, Inc.

Indirect Transmission

Indirect transmission involves the passage of infectious material from a reservoir or source to an intermediate agent and then to a host. The intermediate agent can be living or nonliving. Vehicleborne transmission is accomplished by food, water, biological products (organs, blood, blood products), and **fomites** (inanimate objects) as, for example, desk surfaces, doorknobs, or escalator rails.

Waterborne transmission is a serious problem throughout the world and is a major cause of death in many developing countries as a result of fecal–oral passage, in which pathogens are transmitted from the feces of one individual to another by hand-to-mouth transfer. Public, semiprivate, and private water supplies must all be carefully monitored for the presence of fecal pathogens.

Water and food can serve as both a reservoir and a transmitter of infectious agents. Because of the potential for an outbreak of waterborne and foodborne illnesses, local departments of public health devote considerable attention to sanitary measures designed to minimize risks; their activities include monitoring food packaging industries, food service establishments, beaches, swimming pools, and certification of food handlers.

Fomites play a significant role in the transmission of infectious agents. The list of fomites is seemingly endless and includes objects in common use, such as doorknobs, telephones, faucets, computer keyboards, and exercise equipment. Toys are fomites and contribute to illness in children wherever the toys are shared. Surgical instruments, medical equipment (for example, catheters, intravenous equipment, and syringes), bedding, and soiled clothing are also fomites. An interesting study involving soiled saris as fomites, conducted in fifty-one slum areas in Dhaka, Bangladesh, revealed a positive correlation between the number of misuses of dirty saris and episodes of childhood diarrhea.

The list of fomites and their role may cause you some concern, but there is at least a partial solution to the problem. The simple act of frequent hand washing has been shown to markedly reduce hand-to-mouth (and nose and eye) infection. Frequent wiping of tabletops and counters with disinfectants is effective and a sign of good hygiene in a sanitation-conscious restaurant. In health and exercise clubs it has become a widespread practice to wipe down exercise equipment after use; it is a sign of bad manners not to do so (FIGURE 4.7).

FIGURE 4.7 Exercise machines can be reservoirs for microbes. You may be exercising to maintain good health, but poor personal hygiene may place others at risk. Wipe down exercise machines after use.

Airborne transmission by **aerosols** is the second type of indirect transmission. Aerosols are suspensions of tiny water particles and fine dust in the air; they are distinct from droplet nuclei, as they are smaller than 4 micrometers, travel more than 1 meter, and are small enough to remain airborne for extended periods. Aerosols cause outbreaks of Q fever, Legionnaires' disease, and psittacosis (from infected birds). The microbes in aerosols may not come directly from humans or animals but may be present in dust particles where they can survive for months. Most hantavirus pulmonary syndrome infections can be traced back to when the victim cleaned out mouse droppings from a dusty place such as a summer cabin. Bacteria and viruses can be disseminated by changing bed linens, sweeping, mopping, and other activities. Hospital personnel are keenly aware of this, as reflected in the practice of using wet mops and damp cloths to wipe surfaces.

The third type of indirect transmission is by vectors, living organisms that transmit microbes from one host to another. The term *vector* is sometimes more broadly used to cover any object that transfers microbes, but this is, strictly speaking, incorrect usage. Ticks, flies, mosquitoes, lice, and fleas are the most common vectors, and they belong to the same biological phylum, the **Arthropoda**, along with lobsters and crabs. (It may be difficult to understand what flies, fleas, ticks, and lobsters have in common—but the edibility of lobsters certainly sets them apart!) Arthropods are invertebrate animals with jointed appendages (*arthro* means joint, as in "arthritis" [inflammation of joints], and *pod* means foot, as in "podiatrist" [foot doctor]). Further, they all have segmented bodies and a hardened exoskeleton. The arthropods are members of the largest phylum and consist of many diverse species that are divided into three subphyla (TABLE 4.5). They are considered to be the most successful of all living animals in terms of the huge number of species and their distribution.

Spiders, ticks, and mites hatch from eggs as six-legged larvae and undergo metamorphosis to eight-legged adults with two body segments and mouth parts adapted for the sucking of blood. Ticks transmit a variety of infectious diseases, including Lyme disease, Rocky Mountain spotted fever, babesiosis, and ehrlichiosis. In addition to their role as vectors, some ticks are important reservoirs because they exhibit **transovarial transmission** (the passage of microbes into their eggs).

Insects are an extremely large group of arthropods with well over one million species. You may think of them as pests because (depending on the species) they bite, eat our crops, are bizarre-looking, or are disgusting because they are associated with uncleanliness. Insects have three body segments (the head, the thorax, and the abdomen) and six legs; some have one or two pairs of wings. Some have mouth parts adapted for puncturing the skin and sucking blood. "Kissing bugs" suck blood from their hosts and are vectors of Chagas disease, endemic in Central and South America.

TABLE 4.5 The Phylum Arthropoda	
Subphylum Chelicerata	Subphylum Hexapodia
Scorpions	Insects (many subgroups)
Chiggers	Flies[a]
Spiders	Fleas[a]
Daddy longlegs	Mosquitoes[a]
Mites[a]	Lice[a]
Horseshoe crabs	Subphylum Myriapodia
Ticks[a]	Centipedes
Subphylum Crustacea	Millipedes
Water fleas	
Isopods	
Fairy shrimp	
Crabs	
Copepods[a]	
Lobsters	
Barnacles	
Shrimp	

[a]Vectors of human disease.

Because many arthropods play a significant role in the cycle of infectious diseases, officials at public health departments know that arthropod control can lead to disease control. Mosquito abatement programs have been carried out on numerous occasions to control malaria. All the lower forty-eight states are threatened by West Nile virus, a mosquito-borne virus, resulting in insecticide spraying to control the mosquito population. Several other viruses that cause encephalitis (brain swelling and other neurological damage) belong to a group called arboviruses and are so named because they are arthropodborne (shortened to "arbo").

Arthropods can be either **mechanical vectors** or **biological vectors**. Mechanical vectors transmit microbes passively on their feet and other body parts; the microbes do not invade, multiply, or develop in the vector. Houseflies, for example, feed on exposed human and animal fecal material and then transfer microbes on their feet to food and eating utensils. Typhoid fever and other gastrointestinal diseases characterized by diarrhea or dysentery may be spread in this way. Covering of human and animal waste to avoid exposure to flies is an obvious answer, but this is not always possible in poverty-stricken areas, under wartime conditions, in refugee camps, and in other circumstances involving large groups of people when it is difficult to maintain good sanitation. Even in the best of circumstances, flies have access to dog feces and can mechanically transmit microbes to kitchen areas. It is disturbing to think that a fly that has just lunched on dog feces in your backyard or in a neighboring park may walk across the chicken salad that you prepare for a picnic. Cockroaches also serve as mechanical vectors; remember this when you see them marching across a kitchen counter. In December 2000 Chinese newspapers reported that Beijing was in the grip of a roach menace. Roaches were invading restaurants, hotels and motels, and even hospitals. Roaches carry more than forty kinds of bacteria, some of which are pathogens. Cheap rundown motels are sometimes referred to as "roach motels."

Biological vectors, unlike mechanical vectors, are necessary components in the life cycles of many infectious disease agents and are required for the multiplication and development of the pathogen; transmission by biological vectors is an active process. As an example, when a mosquito picks up the malaria plasmodium parasite while taking a blood meal from an infected person, the parasites are not at an infective stage. Further development in the mosquito's body results in parasites that are now infective for hosts. Depending on the particular vector, parasites may be carried in the saliva and injected into the tissue while biting; other vectors have the nasty habit of regurgitating infectious secretions into and around the bite, and others defecate infectious material onto the bite area. Itching usually results, and scratching facilitates entry of the parasite. Mosquitoes, fleas, lice, and ticks are common biological vectors.

Mosquitoes can rightly be considered as "public health enemy number one" based on their transmission of bacterial, viral, protozoal, and worm diseases (TABLE 4.6). Ticks are not only significant vectors of disease but also are a direct source of disease. Tick paralysis is an example and is characterized by ascending flaccid paralysis resulting from a toxin in tick saliva; the paralysis usually disappears within several days. A 2006 CDC report cited and described a cluster of four cases in Colorado during May 26–31, 2006. Ticks populations depend largely on the number of deer in the area; more deer per square mile means more ticks.

TABLE 4.6 Human Mosquito-borne Disease

Disease	Genus of Mosquito	Genus of Microbe
Caused by viruses		
Eastern equine encephalitis	*Culex, Coquillettidia, Aedes*	Alphavirus
Japanese encephalitis	*Culex*	Flavivirus
La Crosse encephalitis	*Ochlerotatus*	Orthobunyaviridae
St. Louis encephalitis	*Culex*	Flavivirus
West Nile virus	*Culex*	Flavivirus (enveloped, icosahedral nucleocapsid ssRNA)
Western equine encephalitis	*Aedes*	Alphavirus
Dengue fever	*Aedes aegypti*	Flavivirus (dengue virus DEN-1, DEN-2, DEN-3, DEN-4)
Rift Valley fever	*Aedes*	Phlebovirus
Yellow fever	*Aedes*	Flavivirus
Chikungunya fever	*Aedes*	Alphavirus
O'nyong-nyong fever	*Anopheles*	Flavivirus
Ross River Virus	*Culex, Aedes*	Alphavirus
Venezuelan Encephalitis	*Culex, Aedes*	Flavivirus
Murrey Valley encephalitis	*Culex*	Flavivirus
Barman Forest	*Culex, Aedes*	Alphavirus
Australian encephalitis	*Culex*	Flavivirus
California encephalitis	*Aedes*	Orthobunyaviridae
Caused by a protozoan		
Malaria	*Anopheles*	Protozoa *Plasmodium vivax, P. falciparum, P. malariae, P. ovale*
Caused by a worm		
Elephantiasis	*Culex, Anopheles, Aedes, Mansonia, Coquillettidia*	*Wuchereria bancrofti, Brugia malayi*

Fleas are the biological vectors of the *Yersinia pestis* bacterium, and rats are the reservoirs. Rats are important reservoirs for other microbial diseases, including Lassa fever and, in the southeastern United States, hantavirus pulmonary syndrome.

Vectorborne infectious diseases are as numerous and varied as are their vectors, and they are emerging and reemerging throughout the world (TABLE 4.7 and FIGURE 4.8). Factors responsible include genetic changes in both vectors and pathogens resulting in resistance to insecticides and drugs, public health policy, funding directed toward emergency response, and societal changes. In the first half of the twentieth century, considerable progress was made in the fight against vector-

TABLE 4.7 Diseases Transmitted by Arthropod Bites

Disease	Distribution	Vector
Protozoal and helminthic diseases		
Filariasis	Central and South America, Africa, Indian subcontinent and other parts of Asia	Mosquitoes
Babesiosis	United States, Europe	Ticks
American trypanosomiasis	South and Central America	Kissing bug
Onchocerciasis	Central America, tropical South America, Africa	Black flies
African trypanosomiasis (sleeping sickness)	West, Central, and East Africa	Tsetse flies
Leishmaniasis	Central and South America, Africa, Indian subcontinent and other parts of Asia, Europe	Sand flies
Viral diseases		
Yellow fever	Tropical South America, Africa	Mosquitoes
Colorado tick fever	United States (Rocky Mountains)	Mosquitoes
Rift Valley fever	Eastern and southern Africa, sub-Saharan Africa, Madagascar	Mosquitoes
West Nile virus disease	Asia, Africa, United States	Mosquitoes
Dengue fever	India, Southeast Asia, Pacific, South America, Caribbean	Mosquitoes
Bacterial diseases		
Plague	Southeast Asia, Central Asia, South America, western North America	Fleas
Relapsing fever	South America, Africa, Asia, western North America	Lice or ticks
Lyme disease	Europe, United States, Australia, Japan	Ticks
Typhus fever (endemic)	Worldwide	Fleas
Typhus fever (epidemic)	Eastern Europe, Asia, Africa, South America	Lice

From U.S. Department of Health and Human Services, Public Health Service, Centers for Disease Control and Prevention, National Center for Infectious Diseases, Division of Quarantine. 1997. *Health Information for International Travel 1996–97.* http://wonder.cdc.gov/wonder/prevguid/p0000475/p0000475.asp.

borne diseases. Most of these diseases were brought under control, and by the 1960s their threat, except in Africa, was greatly diminished (TABLE 4.8). In fact, malaria was eliminated from many countries of the world. However, no country is immune to the potential threat and spread of vectorborne microbial diseases. A case in point is West Nile virus, which emerged in 1999 in the state of New York. This was the first appearance of the virus in the United States; it had been previously reported only in Africa and Asia.

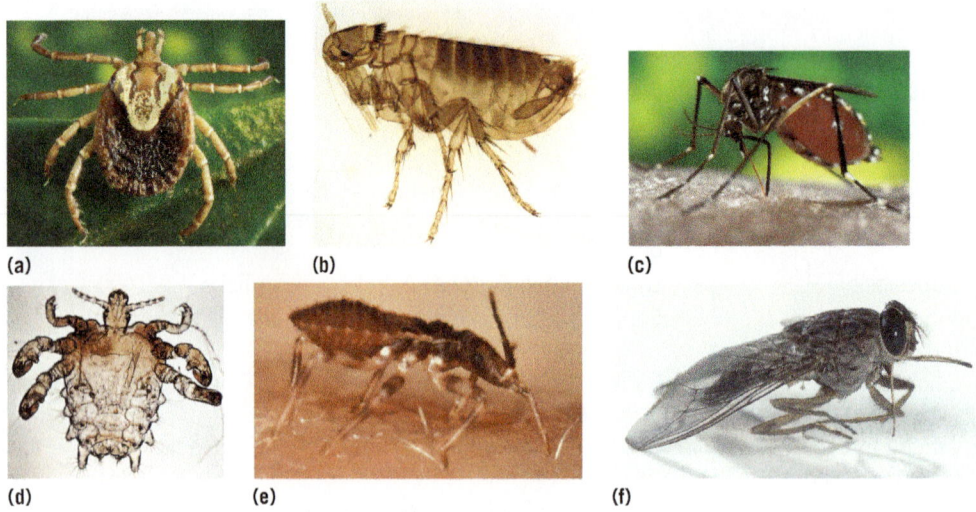

FIGURE 4.8 A "bug" parade: **(a)** tick. Courtesy of James Gathany/CDC. **(b)** flea. Courtesy of John Montenieri/CDC. **(c)** mosquito. Courtesy of James Gathany/CDC. **(d)** louse. Courtesy of WHO/CDC. **(e)** kissing bug. Courtesy of WHO/CDC. **(f)** tsetse fly. Courtesy of Peggy Greb/USDA ARS.

In 1989, in response to the growing problem of vectorborne diseases, the CDC established what is now known as the Division of Vectorborne Infectious Diseases, presently located in Fort Collins, Colorado. The Division is responsible for information, surveillance, prevention, and control of vectorborne diseases. The division is charged with the investigation of national and international epidemics of bacterial and viral diseases transmitted to humans by arthropods, primarily mosquitoes, ticks, and fleas. To prevent and control these diseases, biologists in the division work with the three populations involved: the pathogen, the host, and the vector.

TABLE 4.8 Successful Vectorborne Disease Control

Disease	Location	Year(s)
Yellow fever	Cuba	1900–1901
Yellow fever	Panama	1904
Yellow fever	Brazil	1932
Anopheles gambiae infestation	Brazil	1938
Anopheles gambiae infestation	Egypt	1942
Louseborne typhus	Italy	1942
Malaria	Sardinia	1946
Yellow fever	Americas	1947–1970
Yellow fever	West Africa	1950–1970
Malaria	Americas	1954–1975
Malaria	Global	1955–1975
Onchocerciasis	West Africa	1974–present
Bancroftian filariasis	South Pacific	1970s
Chagas disease	South America	1991–present

Reproduced from Duane J. Gubler and CDC, *Emerging Infectious Diseases*, 4 (1998): 442–450.

Portals of Entry

The next step in the cycle of disease involves access into (or onto) the body through **portals of entry**. Some microbes have a single portal, but others have more than one. Body orifices (openings to the outside), including the mouth, nose, ears, eyes, anus, urethra, and vagina, and penetration of the skin make it possible for microbes to gain access. To some extent, human behavior influences the portal of entry. The most common site of entry for sexually transmitted diseases is the urethra in males and the vagina in females, but the throat and the rectum may also serve for entry. The portal of entry is an important consideration in the outcome of host–parasite interactions. Bubonic plague results from the bite of a plague-infected flea, but if the *Y. pestis* bacteria gains entry into the lungs through the respiratory tract, the result is the more lethal pneumonic plague. Anthrax, also a bacterial disease, is another example. There are three varieties of anthrax: Cutaneous anthrax results when the skin is the portal of entry, gastrointestinal anthrax occurs as the result of oral ingestion of the bacteria, and inhalation anthrax is the result of the organisms' entering through the respiratory tract. TABLE 4.9 and FIGURE 4.9 summarize the portals of entry by anatomical sites.

TABLE 4.9 Infectious Disease Cycle: Portals of Entry and Exit

	Examples of Disease or Microbe
Portals of entry	
Mucous membranes	
Respiratory tract	*Streptococcus pneumoniae,* tuberculosis, Legionnaires' disease, influenza, hantavirus, common cold
Gastrointestinal tract	Cholera, salmonellosis, *E. coli,* rotavirus, poliomyelitis, guinea worm disease, giardiasis
Urogenital tract	Gonorrhea, chlamydia, AIDS, genital warts
Skin (hair follicles, sebaceous glands, wounds, arthropod bites)	Boils, abscesses, cutaneous anthrax, rabies, warts, hookworm, schistosomiasis, malaria
Blood (transfusion, blood products, arthropod bites, placental transfer)	Congenital syphilis, AIDS, German measles, toxoplasmosis, Chagas disease
Portals of exit	
Respiratory tract	Tuberculosis, Legionnaires' disease, influenza, common cold
Gastrointestinal tract	Cholera, salmonellosis, rotavirus, poliomyelitis, hookworm, guinea worm disease
Urogenital tract	Gonorrhea, chlamydia, HIV, schistosomiasis, genital herpes
Skin	Impetigo, boils, abscesses, warts, cold scores, fever blisters, guinea worm disease
Blood (transfusion, blood products, Arthropod bites, placental transfer)	Congenital syphilis, toxoplasmosis, HIV, Rubella, malaria

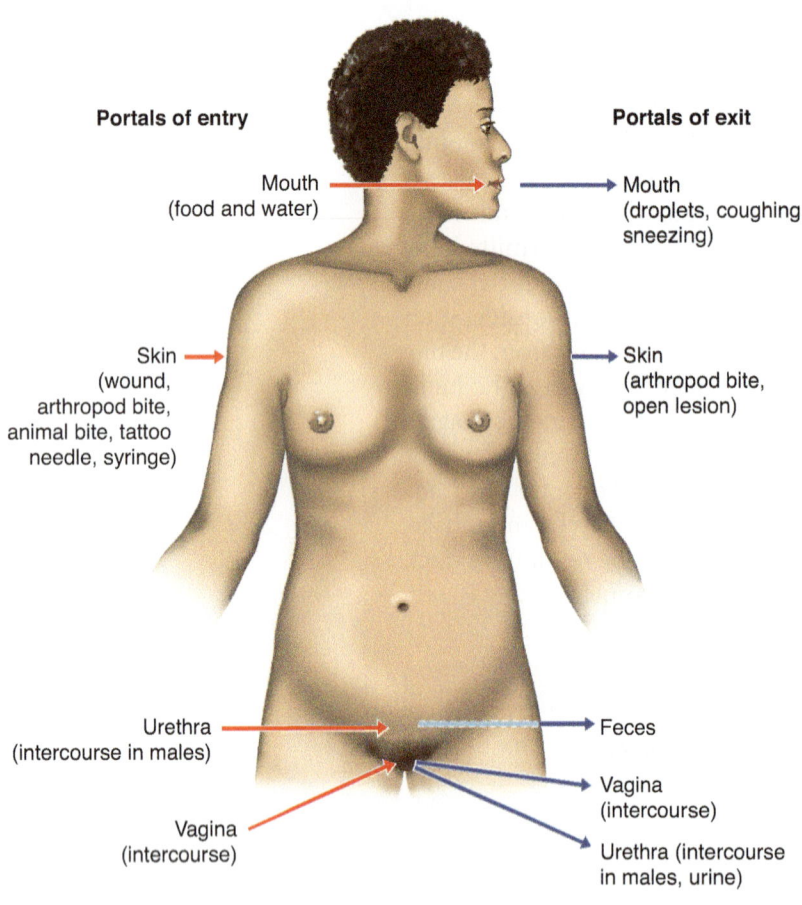

Portals of entry

Mouth
(food and water)

Skin
(wound,
arthropod bite,
animal bite, tattoo
needle, syringe)

Urethra
(intercourse in males)

Vagina
(intercourse)

Portals of exit

Mouth
(droplets, coughing,
sneezing)

Skin
(arthropod bite,
open lesion)

Feces

Vagina
(intercourse)

Urethra (intercourse
in males, urine)

FIGURE 4.9 Portals of entry and portals of exit.

Portals of Exit

Once microbes have gained access into the body, whether or not disease results is determined by the interaction between the number of pathogens and their virulence and the immune system of the host. To complete the cycle of infectious disease and to allow the spread of disease into the community, pathogens require a **portal of exit** (Table 4.9 and Figure 4.9). In some cases the portal of exit relates to the area of the body that is infected. This is particularly true for organisms that cause diseases of the respiratory tract (such as colds and influenza). On the other hand, this is not always the case. For example, the spirochete that causes syphilis uses the urogenital tract as the portal of exit but can invade the skin and nervous system. The eggs of some disease-producing worms exit the body in fecal material, survive in soil, and remain infectious for long periods of time. HIV exits the body through semen and vaginal discharges as well as through the blood. Arthropodborne diseases enter the body through the bites of insects, and these insects also serve as avenues of exit. A mosquito biting an individual with malaria will pick up the parasite.

■ Nosocomial (Hospital-Acquired) Infections

People are hospitalized because they are ill and require treatment beyond what home care can provide. Ironically, while in the hospital they have a 5% to 15% increased risk for contracting an infectious disease. **Nosocomial infections** are infections acquired by patients during their hospital stay or during their confinement in other long-term health care facilities; infections acquired by hospital personnel are also considered nosocomial. Based on the number of hospital admissions, estimates are that two million to four million cases of nosocomial infections, resulting in twenty thousand to forty thousand deaths, occur each year in the United States. The infections account for about 50% of all the major complications of hospitalization.

■ Hospital Environment as a Source of Nosocomial Infections

What are the factors unique to the hospital environment that place patients and hospital staff at an increased risk for acquiring infections? To begin with, the patient population consists of ill individuals who may have a compromised (weakened) immune system. A weak immune system increases a patient's susceptibility to pathogens and opportunistic microbes, including their own normal flora. Antibiotics are heavily used, but also misused, in hospitals to treat or to prevent infections, fostering the development of antibiotic-resistant strains of bacteria. Drugs to purposely suppress the immune system (as in cancer therapy and organ transplantation), prolonged bed rest, and restrictive diets are necessary components of treatment but are traumatic to the body and counterproductive to the maintenance of a healthy immune system.

Diagnostic and treatment protocols frequently involve extensive surgery and the use of invasive procedures, including the insertion of catheters into the urethra, swallowing of tubes, insertion of needles into veins for intravenous therapy, and insertion of nasal tubes. Thermometers, bedpans, urinals, eating utensils, and night table surfaces are only a few of the many fomites that pose potential risk. Hence, the equipment and devices involved in patient care contribute to transmission. The hospital staff, including physicians, nurses, laboratory technicians, and maintenance workers, may unwittingly (and carelessly) transmit microbes from patient to patient; some may be healthy carriers.

All the factors involved in the cycle of infectious diseases are present in a concentrated way in hospitals and in long-term health care facilities, establishing these environments as reservoirs of pathogens. A relatively small number of bacterial species are responsible for most nosocomial infections, but these are species common to the environment. Some sites in the body are more prone to nosocomial infections than others; the urinary tract is the most susceptible, followed by surgical sites and the respiratory tract (TABLE 4.10). Despite the risk of nosocomial infection, be assured that the advances in medicine far outweigh the risks of hospitalization.

Control Measures

Nosocomial infections are a serious problem in hospitals and other medical facilities in terms of mortality, morbidity, and financial burden, and every hospital has strategies of prevention and control. The fact that the frequency and spectrum of antibiotic-resistant organisms are on the rise contributes to the problem. All hospitals are required to have an infection control officer and an infection control committee to maintain accreditation by the American Hospital Association.

TABLE 4.10 Body Site Distribution of Nosocomial Infections	
Site	**Percent of all Nosocomial Infections**
Urinary tract (urinary catheterization)	~50
Surgical site (intestinal surgery, joint replacement surgery)	~25
Lower respiratory tract (respirators and other breathing aids)	~12
Bacteremia (blood infection)	~6
Other (including skin)	~7

Hospitals spend considerable time and money to minimize the possibility of microbial contamination in all aspects of the hospital environment. The infection control officer is responsible for training of hospital personnel in basic infection control procedures including isolation procedures, proper techniques of disinfection and sterilization, and the surveillance and reporting of cases of infectious diseases in both patients and staff. The infection control officer and the infection control committee are also responsible for insect control, good housekeeping, and safe practices for the disposal of feces, urine, bandages, dressings, and other potentially contaminated materials.

Education emphasizing the importance of the simple act of hand washing is vital. (Semmelweis talked about hand washing 160 years ago!) Numerous studies have demonstrated that this single simple procedure is the most important practice in minimizing nosocomial infections. In some studies shockingly low rates (well under 50%) of hand washing by health care workers, including physicians and nurses, have been revealed.

■ Epidemiology of Fear

The fear of epidemics can reach epidemic proportions, but the threat of infectious disease, according to some experts, is out of proportion. *Escherichia coli* outbreaks, avian flu, severe acute respiratory syndrome (i.e., SARS), and West Nile virus have captured the public's attention and led to an explosion of television programs and popular books. The threat of infectious disease, according to some experts, is out of proportion. West Nile virus emerged in the Western Hemisphere for the first time in the summer of 1999 in New York State and caused illness in more than sixty people and the death of seven. By the summer of 2000, infected birds were detected in Massachusetts, Connecticut, and Rhode Island, generating concern among public health officials about an epidemic of fear. As a Massachusetts Department of Health spokeswoman stated, "The message we're trying to get out is to stop people from panicking. West Nile virus is not a major public health threat. It's something people should be aware of, and take precautions, but not let it interrupt their summer."

In his first inaugural address, on March 4, 1933, President Franklin Delano Roosevelt spoke eloquently of the danger of fear. His often quoted words were, "So first of all let me assert my firm belief that the only thing we have to fear is fear itself—nameless, unreasoning, unjustified terror which paralyzes needed efforts to convert retreat into advance." The take-home message is that awareness, surveillance, common sense precautions, and calmness are paramount. Fear can be paralyzing.

■ Overview

Epidemiologists classify disease as sporadic, endemic, epidemic, or pandemic, depending on its frequency and distribution. These categories are not absolute; a particular disease can slide from one classification to another. Common-source epidemics arise from contact with a single contaminant, resulting in a large number of people becoming ill suddenly; the disease peaks rapidly. Propagated epidemics are characterized by direct person-to-person (horizontal) transmission, a gradual rise in the number of infected individuals, and a slow decline.

A chain of linked factors is required for infectious diseases to exist and to spread through a population. These factors are reservoirs of disease, transmission, portals of entry, and portals of exit. Understanding the characteristics of microbes and the diseases they cause is necessary to break the cycle somewhere along its path. The reservoir is the site where microbes exist in nature and from which they can be spread. Active carriers, healthy carriers, and chronic carriers are reservoirs, as are wild and domestic animals. Nonliving reservoirs include contaminated water, food, and soil.

Transmission is the bridge between reservoir and portal of entry. Person-to-person contact is the most common type of horizontal direct transmission and allows for the immediate transfer of microbes. Vertical transmission is another type of direct transmission and is categorized by the passage of pathogens from mother to offspring across the placenta, in the birth canal during delivery, or in breast milk. In direct transmission there are no intermediaries. Indirect transmission involves the passage of materials from a reservoir or source to an intermediate agent and then to a host. The intermediate agent can be nonliving or living. Water, food, fomites, and aerosols are significant nonliving vehicles of indirect transmission. Vectors are living organisms that transmit microbes from one host to another. Mechanical vectors passively transfer microbes on their feet or other body parts, whereas biological vectors are required for the multiplication and development of the pathogen within the vector.

Portals of entry are the next consideration in the cycle of disease. Some microbes have a single preferred portal of entry into the body, whereas others have more than one. Body orifices, including the mouth, nose, ears, eyes, anus, urethra, and vagina, are portals of entry; the skin can be penetrated and is another portal of entry.

For the cycle of disease to continue in a population, microbes must exit from the body. In many cases the portals of entry and the portals of exit are the same.

Hospitals and long-term health care facilities are hotbeds of infection for patients. Hospital-acquired infections are called nosocomial infections and account for about 50% of all the major complications of hospitalization.

The public should not be paralyzed by the fear of infection. Awareness, surveillance, common sense precautions, and calmness are the best preventive measures.

Self-Evaluation

PART I: Choose the single best answer.

1. The breakup of the Soviet Union ushered in an unusually high number of cases of diphtheria from about 1990 to1995. Which term best characterized the situation?

   a. endemic **b.** epidemic **c.** pandemic **d.** herd

2. A common-source outbreak would most likely be attributed to

   a. airborne source **b.** change in vector distribution **c.** person-to-person contact **d.** water supply

3. Chicken pox in the United States is best described as
   a. sporadic **b.** endemic **c.** zoonotic **d.** pandemic
4. Which one of the following is not an insect?
   a. fly **b.** tick **c.** flea **d.** mosquito
5. Vertical transmission is possible except in the case of
   a. gonorrhea **b.** AIDS **c.** influenza **d.** chicken pox

PART II: Fill in the blank.

1. A worldwide outbreak of a disease is called a _____ .
2. Name a disease transmitted by an arthropod vector, and name the vector.
   _____
3. Only one disease has been eradicated. Name this disease. _____
4. Give an example of indirect contact transmission (fomites) of disease.
   _____
5. Which body site is most susceptible to nosocomial infection? _____

PART III: Answer the following.

1. Distinguish between vertical and horizontal transmission. Give examples of each.
2. What is meant by zoonoses? What is a common zoonosis in the northeastern United States? Give three examples of zoonoses.
3. Distinguish between biological vectors and mechanical vectors. Give two examples of each.

CHAPTER 5

Non-Communicable Diseases

By the end of this chapter the student will be able to:

- describe the burden of non-communicable diseases on mortality and morbidity in the United States.
- describe the epidemiological transition and the current distribution of disease in developed and developing countries.
- describe the ideal criteria for a screening program.
- explain the multiple risk factor intervention approach to control of a non-communicable disease.
- describe the meaning of cost-effectiveness.
- describe several ways that genetic interventions can affect the burden of non-communicable diseases.
- describe ways that population interventions can be combined with individual interventions to more effectively reduce the burden of non-communicable diseases.

Sasha didn't want to think about the possibility of breast cancer, but as she turned 50 she agreed to have a mammography which, as she feared, was positive or "suspicious," as her doctor put it. Waiting for the results of the follow-up biopsy was the worst part, but the relief she felt when the results were negative brought tears of joy to her and her family. Then she wondered: is it common to have a positive mammography when no cancer is present?

The first sign of Michael's coronary heart disease was his heart attack. Looking back, he had been at high risk for many years since he smoked, had high blood pressure, and high cholesterol. His lack of exercise and obesity only made the situation worse. Michael asked: what are the risk factors for coronary heart disease and what can be done to identify and address these factors for himself and his family?

John's knee injury from skiing continued to produce swelling and pain, greatly limiting his activities. His physician informed him that the standard procedure today is to look inside with a flexible scope and do any surgery that is needed through the scope. It's simpler, cheaper, and does not even require hospitalization. "We call it 'cost-effective,' " his doctor said. John wondered: what does cost effective really mean?

Jennifer and her husband George were tested for the cystic fibrosis gene and both were found to have it. Cystic fibrosis causes chronic lung infections and greatly shortens the length of life. They now ask: what does this mean for our chances of having a child with cystic fibrosis? Can we find out whether our child has cystic fibrosis early in pregnancy?

Fred's condition deteriorated slowly, but persistently. He just couldn't remember anything and repeated himself endlessly. The medications helped for a short time, but before long he didn't recognize his family and couldn't take care of himself. The diagnosis was Alzheimer's and he was not alone. Almost everyone in the nursing home seemed to be affected. No one seems to understand the cause of Alzheimer's disease. The family asked: what else can be done, not only for Fred, but for those who come after Fred?

Alcohol use is widespread on your campus. You don't see it as a problem as long as you walk home or have a designated driver. Your mind changes one day after you hear about a classmate who nearly died from alcohol poisoning as a result of binge drinking. You ask yourself: what should be done on my campus to address binge drinking?

Each of these scenarios represents one of the approaches to non-communicable diseases that we will examine in this chapter.

WHAT IS THE BURDEN OF NON-COMMUNICABLE DISEASE?

Non-communicable disease represents a wide range of diseases from cardiovascular disease, to cancers, to depression, Alzheimer's, and chronic arthritis. Together, they represent the majority of causes of death and disability in most developed countries. Today, cardiovascular diseases and cancer alone each represent nearly 25 percent of the causes of death as reflected on death certificates in the United States.

The impact of non-communicable diseases on death only reflects part of the impact. Chronic disabilities, largely due to non-communicable diseases, are now the most rapidly growing component of morbidity in most developing as well as developed countries. As populations age, non-communicable diseases increase in frequency. The presence of two or more chronic diseases makes progressive disability particularly likely. The consequences of the rapidly growing pattern of disability due to non-communicable diseases has enormous economic implications. The great increase in direct costs for health care is in part due to the increased burden of non-communicable diseases. The impact extends beyond healthcare costs as it affects the quality of life and may limit the ability of those who wish to work to continue to do so.

Non-communicable diseases have not always dominated the types of diseases which impact a society. Box 5-1 discusses the **epidemiological transition**[1] and provides a perspective on where we stand today.

There are a wide range of preventive, curative, and rehabilitative approaches to non-communicable diseases. However, there are a limited number of basic strategies being used that are part of the population health approach including:

- screening for early detection and treatment of disease
- multiple risk factor interventions
- identification of cost effective treatments
- genetics counseling and intervention
- research

We will take a look at each of these approaches. Finally, we will see how many of these approaches can be combined using the population health approach. Let us begin with what we call **screening** for disease.

BOX 5-1   The Epidemiological Transition and Non-Communicable Diseases.

Disease patterns have not always been the same and will continue to evolve. To gain a big picture understanding of this process of change, it is useful to understand the concept known as the epidemiological or public health transition.

The epidemiological transition describes the changing pattern of disease that has been seen in many countries as they have experienced social and economic development. Its central message is that prior to social and economic development, communicable diseases—or microbial agents using the term from actual causes—represent the dominant cause of disease and disability. In countries in early stages of development, infections are a key cause of mortality either directly or indirectly. For instance, in undeveloped countries maternal and perinatal conditions, as well as nutritional disorders, are often identified as the causes of death. Microbial agents play a key role in maternal and perinatal deaths, as well as deaths ascribed to nutrition. Most maternal deaths are due to infection not necessarily transmitted from others, but related to exposure to microbial agents at the time of birth and in the early postpartum period.

Similarly, most deaths among young children in undeveloped countries are related directly or indirectly to infection. Inadequate nutrition predisposes children to infection and interferes with their ability to fight off infection when it does occur. Many of the deaths among children with malnutrition are related to acute infections, especially acute infectious diarrhea and acute respiratory infections.

As social and economic development progresses, non-communicable diseases including cardiovascular diseases, diabetes, cancers, chronic respiratory ailments, and neuropsychiatric diseases, such as depression and Alzheimer's, predominate as the causes of disability and death. Depression is rapidly becoming one of the major causes of disability and the World Health Organization (WHO) estimates that it will produce more disability than any other single condition in coming years. In addition, illicit drug use has become a major cause of death among the young and includes not only illegal use of drugs, but also abuse of prescription drugs.

In much of the developing world, the same basic patterns are occurring in the developing regions within these countries. Often, earlier distributions of disease dominated by communicable diseases coexist with patterns of non-communicable disease typical of developed countries. Thus, is it not unusual to find that malnutrition and obesity are often present side-by-side in the same developing country.

The epidemiologic transition does not imply that once countries reach the stage where non-communicable diseases dominate that this pattern will persist indefinitely. Newly-emerging diseases, such as HIV/AIDS, pandemic flu, and drug-resistant bacterial infections, raise the possibility that communicable diseases will once again dominate the pattern of disease and death in developed countries.

HOW CAN SCREENING FOR DISEASE ADDRESS THE BURDEN OF NON-COMMUNICABLE DISEASES?

Screening for disease implies the use of tests on individuals who do not have symptoms of a specific disease. These individuals are **asymptomatic**. This implies that he or she does not have symptoms related to the disease being investigated. He or she may have symptoms of other diseases. Screening for disease can result in detection of disease at an early stage under the assumption that early detection will allow for treatment that will improve outcome. Screening has been successful for a range of non-communicable diseases including breast cancer and colon cancer, as well as childhood conditions, including vision and hearing impairments. In all of these conditions, screening has resulted in reduced disability and/or deaths. Not all non-communicable diseases, however, are good candidates for screening and in some cases screening programs have yet to be devised and studied for some non-communicable diseases for which early detection could be useful.

Four criteria need to be fulfilled for an ideal screening program.[2] While few, if any, health conditions completely fulfill all four requirements, these criteria provide a standard against which to judge the potential of a screening program. These criteria are:

1. The disease produces substantial death and/or disability.
2. Early detection is possible and improves outcome.
3. There is a feasible testing strategy for screening.
4. Screening is acceptable in terms of harms, costs, and patient acceptance.

The first criterion is perhaps the easiest to evaluate. Conditions, such as breast and colon cancer, result in substantial death and disability rates. Breast cancer is the second-most common cancer in terms of causes of death and is the most common cancer-related cause of death among women in their 50s. Colon cancer is among the most common causes of cancer death in both men and women. Childhood conditions, such as hearing loss and visual impairment, are not always obvious, however they cause considerable disability.

Determining whether early detection is possible and will improve outcomes is not as easy as it might appear. Screening may result in early detection, but if effective treatment is not available it may merely alert the clinician and the patient to the disease at an earlier point in time without offering hope of an improved outcome. Screening cigarette smokers for lung cancer using X-rays would seem reasonable because lung cancer is the number one cancer killer of both men and women. However, X-ray screening of smokers has been beneficial only in terms of early detection. By the time lung cancer can be seen via chest X-ray, it is already too late to cure. This early detection without improved outcome is called **lead-time bias**.[a]

As indicated in the third criteria, in order to implement a successful screening program, there must be a feasible testing strategy.[b] This usually requires identification of a high-risk population. It also requires a strategy for using two or more tests to distinguish what are called **false positives** and **false negatives** from those who truly have and do not have the disease. False positives are individuals who have positive results on a screening test but do not turn out to have the disease. Similarly, false negatives are those who have negative results on the screening test but turn out to have the disease.

How can we develop feasible testing strategies?[3] To understand the need for and use of feasible testing strategies, it is important to recognize that screening for diseases is usually conducted on groups that are at an increased risk for the condition. For instance, screening men and women for colon cancer and women for breast cancer is often conducted on people aged 50 years and older. This type of group is considered high-risk usually with a chance or risk of having the disease being 1 percent or more. Use of high-risk groups like these allows tests that are less than perfect to serve as initial screening tests.

For instance, mammography has a substantial number of false positives and false negatives. A 50-year-old woman with a positive mammography has only about a 10 to 15 percent chance of having breast cancer. That is, most of the initial positive results will turn out to be false positives.[c]

Therefore, screening for diseases such as breast cancer almost always requires two or more tests. These tests need to be combined using a testing strategy. The most commonly used testing strategy is called **sequential testing** or **two-stage testing**. This approach implies that an initial screening test is followed by one or more definitive or diagnostic test. Sequential testing

---

[a] The concept of lead time implies that screening produces an earlier diagnosis that may be effectively used to intervene prior to diagnosis without screening.

[b] A prerequisite to use of tests in screening and other situations is establishing the cut-off line that differentiates a positive test and a negative test. Often this is done by using what is called a **reference interval**. This range is often established by utilize populations that are believed to be free of a disease. The central 95 percent of the range of values (or the mean plus or minus two standard deviations) for this population is then used as the reference interval or range of normal. This approach has a number of limitations including equating existing levels on a test with desirable levels. When data is available on the desirable levels, it is preferable to utilize these levels to establish a reference interval and to define a positive and negative test result. Establishing the desirable level for a test result requires long term follow-up. For such measurements as blood pressure, cholesterol, and fasting blood sugar, desirable levels are now available.

[c] This assumes a 1 percent pretest probability and a **sensitivity** and **specificity** of 90 percent. A moderately accurate test, such as a mammography, may have a sensitivity of about 90 percent and a similar specificity. Sensitivity implies that the test is about 90 percent accurate in the presence of disease (i.e., present in disease) while specificity tells us the mammography is about 90 percent accurate in the absence of disease (i.e., negative in health).

is used in breast cancer, hearing and vision testing, and most other forms of screening for non-communicable diseases. It is generally the most cost-effective form of screening because only one negative test is needed to rule out the disease.

Sequential testing by definition misses those who have false negative results because a negative test occurs and the testing process is over at least for the immediate future. Thus, a testing strategy needs to consider how to detect those missed by screening. We need to ask: is there a need for repeat screening and if so, when should it occur?[d]

Finally, an ideal screening test should be acceptable in terms of harms, costs and patient acceptance. Harms must be judged by looking at the entire testing strategy—not just the initial test. Physical examination, blood tests, and urine tests often are used as initial screening tests. These tests are virtually harmless. The real question is: what needs to be done if the initial test is positive? If invasive tests such as catheterization or surgery are required, the overall testing strategy may present substantial potential harms.

Screenings and diagnostic tests themselves can be quite costly. In addition, costs are related to the length of time between testing. Testing every year will be far more costly than testing every five or ten years. The frequency of testing depends on the speed at which the disease develops and progresses, as well as the number of people who can be expected to be missed on the initial test. Mammographic screening is traditionally conducted every year because breast cancer can develop and spread rapidly. In the case of colon cancer, however, longer periods between testing are acceptable because the disease is much slower to develop. Thus, cost considerations may be taken into account when choosing between technologies and when setting the interval between screenings.[e]

Finally, patient acceptance is key to successful screening. Many screening strategies present little problem with patient acceptance. However, colon cancer screening has had its challenges with patience acceptance because many consider it an invasive and uncomfortable procedure. Far less then half the people who qualify for screening based upon current recommendations currently pursue and receive colon cancer screening. This contrasts dramatically with mammography where a substantial majority now receives the recommended screening.

The screening tests that completely fulfill these ideal criteria are few and many more are successfully used despite not fulfilling all these criteria. Screening may still be useful as long as we are aware of its limitations and prepared to accept its inherent problems. Table 5-1 illustrates how commonly-used screening tests for risk factors for cardiovascular disease and common cancers perform based upon the four criteria we have outlined.

These criteria do help identify types of screening that should not be done. In general, we do not screen for disease when early detection does not improve outcome. We do not screen for rare diseases, such as many types of cancer, especially when the available tests are only moderately accurate. Finally, we do not screen for diseases when the testing strategy produces substantial harms. Screening for disease is not the only population health approach that can be used to address the burden of non-communicable disease. Multiple risk factor reduction is a second strategy that we will examine.

HOW CAN IDENTIFICATION AND TREATMENT OF MULTIPLE RISK FACTORS BE USED TO ADDRESS THE BURDEN OF NON-COMMUNICABLE DISEASE?

As we have seen, the concept of risk factors is fundamental to the work of public health. Risk factors ranging from high levels of blood pressure and LDL cholesterol to multiple sexual partners and anal intercourse help us identify groups that are most likely to develop a disease. Evidence-based recommendations often focus on addressing risk factors and implementation efforts often address the best way(s) to target high-risk groups. Thus, identifying and reducing risk factors is an inherent part of the population health approach to non-communicable diseases.

A special form of intervention aimed at risk factors is called **multiple risk factor reduction**. As the name implies, this strategy intervenes simultaneously in a series of risk factors all of which contribute to a particular outcome, such as cardiovascular disease or lung cancer. Multiple risk factor reduction is most effective when there are constellations or groups of risk factors that cluster together in definable groups of people. It may also be useful when the presence of two or more risk factor increases the risk more than would be expected by adding together the impact of each risk factor.

The success of the last half century in addressing coronary artery disease exemplifies multiple risk factor reduction. Box 5-2 discusses the impact of the strategy on coronary

---

[d] A sequential testing strategy also requires a decision on the order of administering the tests. Issues of cost and safety are often the overriding consideration in determining which test to use first and which to use to confirm an initial positive test. At times, a testing strategy known as **simultaneous testing** or **parallel testing** is used. In this scenario, two tests are used initially if one test can be expected to detect one type of disease and the other test can be expected to detect a different type of disease. Traditionally, flexible sigmoidoscopy, which examines the lower approximately 35 cm of the colon, has been used along with tests for occult blood in the stool. Tests for occult blood attempt to screen for cancer in the large section of colon proximal to the sigmoid region. Using both of these tests has been shown to be more accurate for screening than use of either test alone because each attempts to find cancer in different anatomical sites.

[e] Today, there is a wide range of methods for screening for colon cancer including colonoscopy, which examines the entire colon, and virtual colonoscopy, which does not require an internal examination. These newer tests are much more costly than sigmoidoscopy and occult stool testing. Which is the most accurate and cost-effective test remains controversial. However, the need for and benefits of screening for colon cancer are widely accepted.

TABLE 5-1 Examples of Screening Tests for Heart Disease and Cancer and Ideal Criteria

	Substantial mortality and/or morbidity	Early detection possible and alters outcome	Screening is feasible (can identify a high-risk population and a testing strategy)	Screening acceptable in terms of harms, costs, and patient acceptance
Hypertension	Contributory cause of strokes, myocardial infarctions, kidney disease	High blood pressure precedes bad outcomes often by decades and effective treatment is available	Test everyone—desirable range has been established	Screening itself is free of harms, low cost, and acceptable to patients Treatments, however, may be complicated and have harms, costs, and side effects
LDL cholesterol	Contributory cause of strokes, myocardial infarctions, and other vascular diseases	Precedes the development of disease by decades and treatment is effective in altering outcome	Test everyone—desirable range has been established	Screening itself is free of harm, low cost, and acceptable to patients Treatment has rare side effects, which can be detected by symptoms and low-cost blood tests
Breast cancer	2nd most common fatal cancer among women and most common for women under 70	Early detection improves outcome	For those 50 and over, combination of mammography and follow-up biopsy shown to be feasible	Harm may occur due to false positives, low risk of harm from radiation, patient acceptance good, but test can be somewhat painful Screening younger women increases costs and false positives
Cervical cancer	If undetected and untreated—may be fatal	Early treatment dramatically reduces the risk of death	Pap smear and follow-up testing have been extremely successful	Pap results in substantial number of false positives New DNA testing may be used to separate true and false positives
Colon cancer	2nd most common fatal cancer in men and third in women	Early detection of polyps reduces development of cancer and early detection of cancer improves chances of survival	Men and women 50 and older, plus those with high risk types of colon disease Options for screening include: fecal occult blood testing, plus flexible sigmoidoscopy, colonoscopy, and virtual colonoscopy	Patient acceptance has been major barrier, small probability of harm from procedure, substantial cost for colonoscopy and virtual colonoscopy

BOX 5-2 Coronary Artery Disease and Multiple Risk Factor Reduction.

An epidemic of coronary artery disease and subsequent heart attacks spread widely through mid-20th century America. Sudden death, especially among men in their 50s and even younger, became commonplace in nearly every neighborhood in suburban America. To better understand this epidemic, which caused nearly half of all deaths in Americans in the 1940s and 1950s, the National Institutes of Health began the Framingham Heart Study in the late 1940s.[4]

In those days, there were only suggestions that cholesterol and hypertension contributed to heart disease and little, if any, recognition that cigarettes played a role. The Framingham Heart Study enrolled a cohort of over 7000 individuals in Framingham, Massachusetts: questioning, examining, and taking blood samples from them every other year to explore a large number of conceivable connections with coronary artery disease—the cause of heart attacks. Now well into its second half-century after thousands of publications and hundreds of thousands of examinations, the Framingham Heart Study continues to follow the children and grandchildren of the original Framingham cohort.

The study has provided us with extensive long-term data on a cohort of individuals. These form the basis for many of the numbers we use to estimate the strength of risk factors for coronary artery disease. It has helped demonstrate not only the risk factors for the disease, but also the protective or resilience factors. The use of aspirin, regular exercise, and modest alcohol consumption have been suggested as protective factors despite the fact that no one ever thought of them in the 1940s.[f]

The Framingham Heart Study demonstrated that high blood pressure preceded strokes and heart attacks by years and often decades. It took the Veterans Administration's randomized clinical trials of the early 1970s to convince the medical and public health communities that high blood pressure needed and benefited from aggressive detection and treatment. Through a truly joint effort by public health and medicine, high blood pressure detection and treatment came to public and professional recognition as a major priority in the 1970s.

The impact of elevated levels of low-density lipoprotein (LDL), the bad cholesterol, were likewise suggested by the Framingham Heart Study, but it was not until the development of a new class of medications called statins in the mid-1980s that treatment of high levels of LDL cholesterol took off. These drugs have been able to achieve remarkable reductions in LDL and equally remarkable reductions in coronary artery disease with only rare side effects. These drugs have been so successful that some countries have made them available over-the-counter. Clinicians are using them more and more aggressively to achieve levels of LDL cholesterol that are less than half those sought a generation ago.[g]

Although diabetes has been treated with insulin since the 1920s and oral treatments beginning after World War II, the treatment of diabetes to prevent its consequences—including coronary artery disease—was not definitely established as effective until the 1990s. Our current understanding of diabetes has come from a series of randomized clinical trials and long-term follow-ups that demonstrate the key role that diabetes can play in diseases of the heart and blood vessels and the impressive role that aggressive treatment can play in reducing the risks of these diseases.

Efforts aimed at early detection and treatment of heart attacks and prevention of second heart attacks through the use of medications have become routine parts of medical practice. Medical procedures, including angioplasty and surgical bypass of diseased coronary arteries, have also been widely used. Widespread availability of defibrillators in public areas is one of the most recent effort to prevent the fatal consequences of coronary artery disease.

Between the 1950s and the early years of the 21st century, the death rate from coronary artery disease has declined by over 50 percent. The impact is even greater among those in their 50s and 60s. Sudden death from coronary artery disease among men in their 50s is now a relatively rare event.

For years medicine and public health professionals debated whether public health and clinical preventive interventions or medical and surgical interventions deserved the lion's share of the credit for these achievements. The evidence suggests that both prevention and treatment have had important impacts.[6] When medicine and public health work together, the public's health is the winner.

---

[f] The data developed for the Surgeon General's Reports on Smoking and Health in the 1960s and beyond, strongly pointed to substantial effects of cigarettes not only on lung disease, but on coronary artery disease as well. In fact, given the large number of deaths from coronary artery disease compared to lung disease, it became evident that in terms of number of deaths the biggest impact of cigarette smoking is on heart disease, not lung disease.

[g] Recent data even suggests that for individuals with levels of LDL cholesterol within the currently accepted range of normal, statins may be beneficial in the presence of evidence of inflammation as measured by a test called C-reactive protein.[5]

artery disease. Multiple risk factor reduction strategies are being attempted for a range of diseases from asthma to diabetes.

Multiple risk factor reduction is most successful when a number of risk factors are at work in the same individual. As we have seen with asthma, factors like indoor and outdoor air pollution, cockroaches and other allergens, and a lack of adherence to medications, tend to occur together and may be most effectively addressed together. Similarly, obesity and lack of exercise tend to reinforce each other often requiring a comprehensive multiple risk factor reduction approach.[h]

Screening for disease and multiple risk factor reduction are key approaches to using testing as part of secondary intervention.[i] The enormous burden of non-communicable disease cannot be totally prevented even by maximizing the use of these strategies. It is important to couple them with cost-effective treatment. Thus, a third population health strategy for addressing the burden of non-communicable disease is to develop cost-effective interventions to treat common diseases.

HOW CAN COST-EFFECTIVE INTERVENTIONS HELP US ADDRESS THE BURDEN OF NON-COMMUNICABLE DISEASES?

Clinicians today have a wide range of interventions to treat disease. Many of these interventions have some impact on the course of a disease. The proliferation of interventions means that it is especially important to identify which provide the greatest benefits at the lowest cost. In order to understand how cost-effective interventions can help address the burden of non-communicable disease, we need to understand what we mean by cost-effective.

Cost-effectiveness is a concept that combines issues of benefits and harms with issues of financial costs. It starts by considering the benefits and harms of an intervention to determine its **net-effectiveness**. Net-effectiveness implies that the benefits are substantially greater than the harms even after the value (or utility), as well as the timing of the harms and benefits, are taken into account. Only after establishing net-effectiveness do we take into account the financial costs.

Cost-effectiveness compares a new intervention to the current or standard intervention. It usually asks: is the additional net-effectiveness of an intervention worth the additional cost? At times it may also require us to ask: is a small loss of net-effectiveness worth the considerable savings in cost? Figure 5-1 is a tool for categorizing interventions in order to analyze their costs and net-effectiveness.

Box 5-3 provides more details on the use of cost-effectiveness analysis.[7]

Preventive interventions often undergo cost-effectiveness analysis. Many interventions, ranging from mammography to most childhood vaccinations to cigarette cessation programs, get high or at least passing grades on cost-effectiveness. However, many widely-used treatment interventions do not or would not meet the current standards of cost-effectiveness. The application of cost-effectiveness criteria to common clinical interventions is considered a population health intervention aimed at getting maximum value for the dollars spent.

The results of cost-effectiveness analysis have already had an impact on a number of common clinical procedures. For instance, cost-effective treatments include: the use of minimally-invasive orthopedic surgery, such as knee surgery; the reduced length of intensive care and hospitalization for coronary artery disease; and the use of home health care for intravenous

[h] In some situations the existence of multiple risk factors does more than add together to produce disease. At times, the existence of two or more factors multiply the risk. In these situations addressing even one of the factors can have a major impact on disease. For instance, it is now well established that asbestos exposure and cigarette smoking multiply the risks of lung cancer. Thus, if the relative risk for cigarettes is 10 and the relative risk for asbestos exposure is 5, then the relative risk if both factors are present is approximately 50. If an individual who has previously been exposed to both risk factors stops smoking cigarettes and the effects of cigarette smoking are immediately and completely reversible, we can expect the relative risk of lung cancer to decline from approximately a fifty-fold increase to a five-fold increase.

[i] The principles of testing discussed here are not limited to screening for disease and identification of risk factors. They are also useful as part of a cost-effective approach to diagnosis of symptomatic diseases. In addition, testing is often used for a range of applications in medicine and public health, including monitoring response to treatment, identifying side effects, identifying genetic predictors of disease, and establishing baseline levels for future testing. Public health applications include the use of environmental testing and testing for disease prevalence.

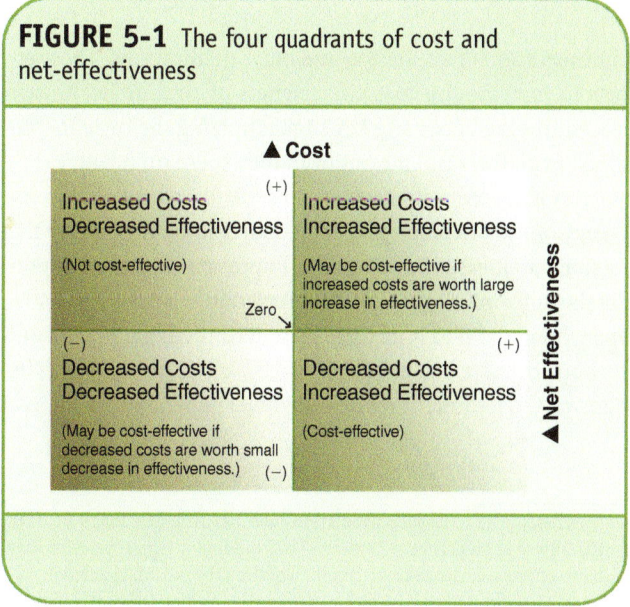

FIGURE 5-1 The four quadrants of cost and net-effectiveness

BOX 5-3   Cost-Effectiveness and Its Calculations.

Cost-effectiveness is often judged by comparing the costs of a new intervention to the cost of the current, standard, or state-of-the-art intervention. A measure known as the **incremental cost-effectiveness ratio** is then obtained. This ratio represents the additional cost relative to the additional net-effectiveness.

Net-effectiveness may measure a diagnosis made, a death prevented, or a disability prevented, etc. To operationize the concept of net-effectiveness requires us to define, measure, and combine the probabilities and utilities of benefits with the probabilities and utilities of harm and take into account the timing of the benefits and the harms. Thus, the process of calculating net-effectiveness can be quite complex.

Similarly, calculating costs can be challenging. Most economists argue that the costs are not limited to the costs of providing the intervention and the current and future medical care, but should also include the cost of transportation, loss of income, and other expenses associate with obtaining health care and being disabled. Thus, calculating a cost-effectiveness ratio has become a complex undertaking.

The criteria for establishing cost effectiveness have changed over time. Most experts in cost-effectiveness prefer the use of a measurement called **quality-adjusted life years** or **QALYs**. QALYs ask about the number of life-years saved by an intervention, rather than the number of lives. Thus, one QALY may be thought of as one year of life at full health compared to immediate death.

In cost-effectiveness analysis, a financial value is usually placed on a QALY reflecting what a society can afford to pay for the average QALY as measured by its per capita gross domestic product (GDP). In the United States, where the GPD is approaching $50,000 there is a general consensus that a QALY currently should be valued at $50,000. Thus, when you hear that a formal cost-effectiveness analysis has shown that an intervention is cost-effective, it generally implies that the addition cost is less than $50,000 per QALY.[j]

The ideal intervention is one in which the cost goes down and the effectiveness goes up. Cost-saving, quality-increasing interventions have a negative incremental cost-effectiveness ratio. That is, QALYs go up while the costs go down. These cost-reducing, QALY-increasing interventions, while highly desirable, are very rare. Usually we need to spend more to get additional QALYs. One example of an intervention that reduces costs and at the same time produces additional QALYs is treatment of hypertension in high risk individuals, such as those with diabetes.[k]

[j] An increase of one QALY may be the result of obtaining small improvements in utility from a large number of people. For instance, if ten people increase their utility from 0.1 to 0.2, the result is an increase of one QALY. The value of a QALY is often set slightly above the average gross domestic product (GDP) or sum of all goods and services produced per person. In the United States, the GDP now exceeds $40,000 per person per year. Thus, setting the value of a QALY at $50,000 reflects how much we can afford to pay rather than strictly reflecting how much we think a QALY is worth.

[k] Another example is the use of influenza vaccine among the elderly and those with chronic disease predisposing them to the complications of influenza. It is important to distinguish these types of interventions from those that reduce the costs, but also reduce the QALYs, because at times both are referred to as cost-saving measures.

administration of antibiotics and other medications. These efforts to increase the cost-effectiveness of routine healthcare procedures are becoming key to maximizing the benefits obtained from the vast amount of money spent on health care.

Applying cost-effectiveness analysis to routine clinical interventions is often coupled with efforts to better predict the outcome of disease and treatment. Improving the ability to predict the outcome of diseases and interventions can help us know when, how, and if to intervene. Improved prediction holds out the hope of increased effectiveness, as well as reduced costs by tailoring the treatment to the individual patient.[l]

In addition to screening, multiple risk factor reduction, and cost-effective intervention using prediction rules, the revolution in genetics has opened up another possible strategy for addressing the burden of non-communicable diseases.

HOW CAN GENETIC COUNSELING AND INTERVENTION BE USED TO ADDRESS THE BURDEN OF CHRONIC DISEASES?

Interventions based upon genetics have been part of medical and public health practice since at least the 1960s, when it was recognized that abnormalities of single genes for such condi-

[l] Research designed to develop **prediction rules** has become a major focus of clinical, as well as public health research. This challenging type of research may be able to improve evidence-based recommendations by providing different recommendations for groups with different prognoses and/or different responses

to treatment. Efforts to improve the effectiveness of breast cancer treatments, for instance, are now focused on testing patients to determine the best type of chemotherapy to use to treat their particular cancer. This approach shows promise of producing greater benefit and less harm at reduced cost.

tions as Tay-Sachs disease (found among Ashkenazi Jews) and sickle cell anemia (found among African-Americans) could be detected by testing potential parents who could then be counseled on the risks associated with childbearing.

It was also recognized that chromosomal abnormalities that produce Down Syndrome, the most commonly-recognized cause of mental retardation, could be detected at an early stage in pregnancy. In addition, certain genetic defects such as phenylketonuria (PKU) can be recognized at birth and relatively simple dietary interventions can prevent the severe retardation of mental development that would otherwise occur.

In light of all this valuable knowledge, today genetic testing and counseling are often offered to prospective parents. Testing for Down Syndrome is a standard part of prenatal care; and testing for a wide range of rare, but serious disorders is a population health intervention. In fact, in most states these tests are legally required soon after birth. The triumph of the human genome project in the early years of the 21st century has sparked interest in expanding the applicability of genetic interventions in medicine and public health. For instance, the gene for cystic fibrosis, the most common genetic disorder among whites in the United States, has been identified and screening of large numbers of couples is now possible. Even among whites without a history of cystic fibrosis, the chance of carrying the gene is about three percent. If both the mother and the father carry the gene, the chances of having a child with cystic fibrosis is 25 percent with each pregnancy. The fetus can be tested for the disease early in pregnancy.

There are a wide range of current and developing applications of genetics including:[8]

- **Genetic prevention**—This approach incorporates efforts to prevent the occurrence of single genes or multiple gene combinations that are likely to produce disease. This includes: expanded use of genetic counseling, prenatal testing, and early abortion or fetal therapies. Knowledge of the human genome holds promise for expanding this approach beyond diseases caused by single-gene defects to diseases that depend on multiple genes. It is important to recognize that diseases that are dependent on multiple genes will be more difficult to predict than resulting from a single-gene.
- **Genetic detection prior to disease**—This approach includes efforts aimed at detection of genetic defects and implementation of early intervention to prevent what is called the **phenotypic expression of genes**. Building on the success of treatment of PKU and other inborn errors of metabolism, risk factors for common diseases, such as high cholesterol, high blood sugar, or obesity might be detected early and aggressively managed during childhood.
- **Gene-environmental protection**—Genetic testing holds out the possibility of defining combinations of genes that identify individuals who are especially likely to develop disease when they experience specific environmental exposures, such as those interactions that occur in occupational settings where workers are exposed to specific chemicals often at low doses. Identification of gene-environment interactions may lead to identification of those who are at high risk if they work in certain occupational settings.
- **Genotypic-based screening for early disease**—Combinations of genes may identify groups that are at high risk of common diseases and that can be targeted for screening. For instance, studies suggest that for certain common cancers, such as those of the breast, prostate, and colon, genetic factors are associated with 30–40 percent of these diseases. Finding predisposing genetic patterns early in life may be useful for identifying those who need earlier or more intensive screening for early detection.

Each of these potential uses of genetics in public health present ethical as well as technological issues. Questions to consider include: Should we identify diseases when little can be done to prevent or treat them? How can we identify genetic risk factors without stigmatizing or putting those with the genes at a disadvantage? Will screening programs be improved by identifying groups at genetic risk of disease or will high risk groups without genetic risk factors unfairly be passed over for screenings?

These and other approaches based on the rapidly accumulating knowledge of genetics are likely to become routine medical and public health strategies for primary, secondary, and tertiary interventions for non-communicable disease in the coming years. Their successful adoption, however, will require careful attention by those in the fields of public health and medicine to ensure that benefits are gained and harms minimized. Despite the enormous advances that have occurred in public health and medicine in recent decades, there is still much to be learned. A final strategy for non-communicable diseases addresses the question: what can we do when highly-effective interventions don't exist?

WHAT CAN WE DO WHEN HIGHLY-EFFECTIVE INTERVENTIONS DON'T EXIST?

Alzheimer's disease reflects the challenge of what to do when the cause of a disease is not known and the treatment is not highly effective.[9] Alzheimer's is among the most rapidly increasing condition among those that we classify as non-communicable diseases. The aging of the population has been and is expected to be associated with many more cases of Alzheimer's, which primarily affects the quality of life with its progressive damage to memory—especially short-term memory.[m]

Today we have limited treatment options for those afflicted with Alzheimer's. Several drugs are available that have modest positive impacts on memory. Efforts to stimulate mental activity through keeping active mentally and physically have also been shown to have positive, yet modest impacts. Public health efforts have encouraged the use of these existing interventions especially when there is evidence that they allow individuals to function on their own or with limited assistance for longer periods of time.

The population health approach to Alzheimer's disease, however, also stresses the need for additional research. Epidemiological research has helped produce the modest advances in preventing progression and treating the symptoms of the disease. A population health approach, however, needs to acknowledge the need for a basic biological understanding of what causes Alzheimer's. The P.E.R.I. process asks us to address the etiology as the basis for evidence-based recommendations and intervention. Thus, the population health approach to Alzheimer's, as with other diseases of unknown etiology, requires us to ask basic questions about the biology of the disease and to learn more about its cause(s). Fortunately, an increasingly sophisticated and well-financed effort is being directed at understanding the etiology of Alzheimer's. We can now have hope and increasing confidence that the epidemic being faced by many of your grandparents will be brought under control in the not-too-distant future.

We have now explored the major population health strategies for addressing non-communicable diseases. These include screening, multiple risk factor reduction, cost-effective treatments, genetic counseling, and more research. A complex problem often requires us to combine many of these approaches.

_____

[m] Not all cases of memory loss or dementia are due to Alzheimer's. Additional causes include: strokes and cerebral vascular disease, chronic alcoholism, thyroid disease, specific infectious diseases (such as syphilis and AIDS), as well as the effects of drugs and a long list of rare diseases. Today, however, Alzheimer's is the most common and the most important cause of memory loss and dementia. We tend to classify a disease as non-communicable unless there is convincing evidence that it can be transmitted or that it is due in large part to environmental exposures or injuries. Despite the fact that we do not yet know the etiology of Alzheimer's disease, it is generally classified as non-communicable.

HOW CAN WE COMBINE STRATEGIES TO ADDRESS COMPLEX PROBLEMS OF NON-COMMUNICABLE DISEASES?

Multiple interventions combining health care, traditional public health approaches, and social interventions are often needed to address the complex problems presented by non-communicable diseases. The combined and integrated use of

---

BOX 5-4   Alcohol Abuse and the Population Health Approach.

Alcohol has been a central feature of American society and American medicine and public health since the early days of the country. It was among the earliest painkillers and was used routinely to allow surgeons to perform amputations during the Civil War and earlier conflicts. The social experiment of alcohol prohibition during the 1920s and early 1930s ended in failure as perceived by a great majority of Americans.

Efforts to control the consequences of alcohol took a new direction after World War II. Americans began to focus on the consequences of the disease, including liver disease, fetal alcohol syndrome, automobile accidents, and intentional and unintentional violence.

Population health interventions became the focus of alcohol control efforts. For instance, taxation of alcohol based upon 1950s legislation raised the price of alcohol enough to substantially reduce consumption. Restrictions on advertising and higher taxes on hard liquor with its greater alcohol content eventually contributed to greater use of beer and wine. Despite the continued growth in alcohol consumption, the number of cases of liver disease and other alcohol-related health problems have declined. In recent years, efforts to alert pregnant women to the health effects of drinking through product labeling and other health communications efforts have had an impact.

The highway safety impacts of alcohol use have led to population health efforts in cooperation with transportation and police departments. Greatly increased police efforts to catch drunk drivers and stripping of the licenses of repeat offenders have become routine and have been attributed to impressive reductions in automotive accidents related to alcohol. Efforts such as the designated driver movement originated by Mothers Against Drunk Drivers (MADD) have demonstrated the often-critical role that private citizens can play in implementing population health interventions.

(continues)

BOX 5-4   continued.

Focusing on high-risk groups, as well as using what we have called "improving-the-average" strategies, has had an important impact. Alcoholics Anonymous (AA) and other peer support groups have focused on encouraging individuals to acknowledge their alcohol problems. These groups often provide important encouragement and support for long-term abstinence.

Medical efforts to control alcohol consumption have been aimed primarily at those with clear evidence of alcohol abuse—often those in need of alcohol withdrawal or "drying out." Drugs are available that provide modest help in controlling an individual's alcohol consumption. Screening for alcohol abuse has become a widespread part of health care. These interventions have been aimed at those with the highest levels of risk. The combination of individual, group, and population interventions has reduced the overall impact of alcohol use without requiring its prohibition. In fact, modest levels of consumption, up to one drink per day for women and two for men, may help protect against coronary artery disease.

The issue of alcohol and public health has not gone away. The focus today has returned to identifying high-risk groups and intervening to prevent bad outcomes. A key risk factor today is binge drinking with its risk of acute alcohol poisoning, as well as unintentional and intentional violence. College students are among the highest risk group. One episode of binge drinking dramatically increases the probability of additional episodes suggesting that intervention strategies are needed to reduce the risk. We've made a great deal of progress controlling the impacts of alcohol, but we clearly have more to do.

multiple interventions is central to the population health approach. Box 5-4 looks at what we can learn about the population health approach to non-communicable diseases from the long history of alcohol use and abuse, as well as the substantial recent success in addressing disease due to alcohol.[10]

We have now taken a look at strategies to control non-communicable diseases that are currently the most common reason for disability and death in most developed countries. Now, let us look at a second category—communicable disease—which has been central to the history of public health and threatens to become central to its future.

Key Words

- Epidemiological transition
- Screening
- Asymptomatic
- Lead-time bias
- Watchful waiting
- False positives
- False negatives
- Reference interval
- Sensitivity

- Specificity
- Sequential testing (or two-stage testing)
- Simultaneous testing (or parallel testing)
- Multiple risk factor reduction
- Cost-effectiveness
- Net-effectiveness
- Prediction rules
- Phenotypic expression of genes

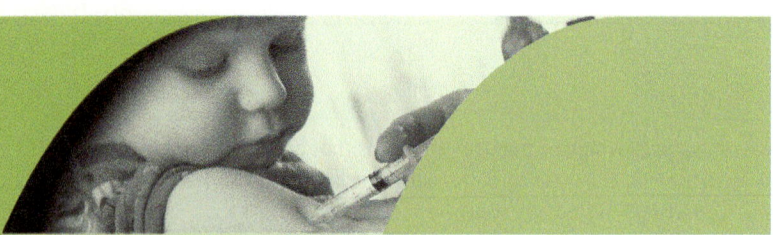

Discussion Questions

Take a look at the questions poised in the following scenarios which were presented at the beginning of this chapter. See now whether you can answer them.

1. *Sasha didn't want to think about the possibility of breast cancer, but as she turned 50 she agreed to have a mammography which, as she feared, was positive or "suspicious," as her doctor put it. Waiting for the results of the follow-up biopsy was the worst part, but the relief she felt when the results were negative brought tears of joy to her and her family. Then she wondered: is it common to have a positive mammography when no cancer is present?*

2. *The first sign of Michael's coronary heart disease was his heart attack. Looking back, he had been at high risk for many years since he smoked, had high blood pressure, and high cholesterol. His lack of exercise and obesity only made the situation worse. Michael asked: what are the risk factors for coronary heart disease and what can be done to identify and address these factors for himself and his family?*

3. *John's knee injury from skiing continued to produce swelling and pain, greatly limiting his activities. His physician informed him that the standard procedure today is to look inside with a flexible scope and do any surgery that is needed through the scope. It's simpler, cheaper, and does not even require hospitalization. "We call it 'cost-effective,' " his doctor said. John wondered: what does cost effective really mean?*

4. *Jennifer and her husband George were tested for the cystic fibrosis gene and both were found to have it. Cystic fibrosis causes chronic lung infections and greatly shortens the length of life. They now ask: what does this mean for our chances of having a child with cystic fibrosis? Can we find out whether our child has cystic fibrosis early in pregnancy?*

5. *Fred's condition deteriorated slowly, but persistently. He just couldn't remember anything and repeated himself endlessly. The medications helped for a short time, but before long he didn't recognize his family and couldn't take care of himself. The diagnosis was Alzheimer's and he was not alone. Almost everyone in the nursing home seemed to be affected. No one seems to understand the cause of Alzheimer's disease. The family asked: what else can be done, not only for Fred, but for those who come after Fred?*

6. *Alcohol use is widespread on your campus. You don't see it as a problem as long as you walk home or have a designated driver. Your mind changes one day after you hear about a classmate who nearly died from alcohol poisoning as a result of binge drinking. You ask yourself: what should be done on my campus to address binge drinking?*

REFERENCES

1. Omran AR. The epidemiologic transition: a theory of the epidemiology of population change. *The Milbank Memorial Fund Quarterly.* 1971; 49(4): 509–538.

2. Riegelman RK. *Studying a Study and Testing a Test: How to Read the Medical Evidence.* Philadelphia: Lippincott, Williams & Wilkins; 2005.

3. Gordis L. *Epidemiology.* 4th ed. Philadelphia: Elsevier Saunders; 2009.

4. Framingham Heart Study. History of the Framingham Heart Study. Available at: http://www.framinghamheartstudy.org/about/history.html. Accessed March 18, 2009.

5. Ridker PM, Danielson E, Foneseca FAH, Genest J, et al. Rosuvastatin to prevent vascular events in men and women with elevated C-reactive protein. *N Engl J Med.* 2008; 259(21): 2195–2207.

6. Critchley J, Capwell S, Unal B. Life-years gained from coronary heart disease mortality reduction in Scotland—prevention or treatment? *J Clin Epi.* 2003; 56(6): 583–590.

7. Gold MR, Siegel JE, Russell LB, Weinstein MC. *Cost-Effectiveness in Health and Medicine.* New York: Oxford University Press; 1996.

8. Khoury MJ, Burke W, Thomson EJ. *Genetics and Public Health in the 21st Century: Using Genetic Information to Improve Health and Prevent Disease.* New York: Oxford University Press; 2000.

9. National Institute on Aging. Alzheimer's Disease Fact Sheet. Available at: http://www.nia.nih.gov/Alzheimers/Publications/adfact.htm. Accessed March 18, 2009.

10. Room R, Babor T, Rehm J. Alcohol and public health. *Lancet.* 2005; 365(9458): 519–530.

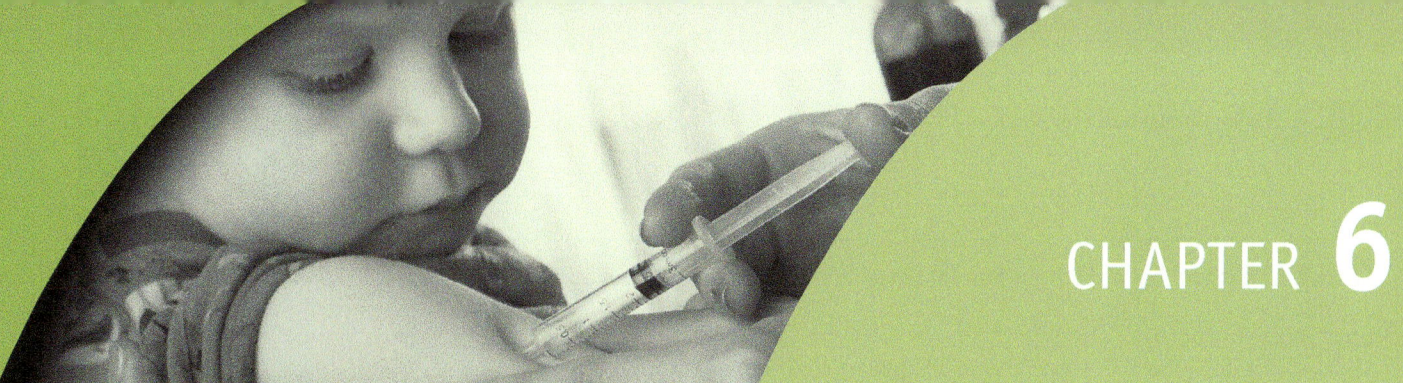

CHAPTER **6**

Social and Behavioral Sciences and Public Health

LEARNING OBJECTIVES

By the end of this chapter the student will be able to:

- explain relationships between the social and behavioral sciences and public health.
- illustrate how socioeconomic status affects health.
- illustrate how culture and religion affects health.
- identify and illustrate the stages in behavioral change that constitute the Stages of Change model.
- identify ways that interventions at the individual level and at the social level can reinforce each other to influence behavioral change.
- explain the principles of social marketing.

You travel to a country in Asia and find that their culture affects most parts of life. From the food they eat and their method of cooking, to their attitudes toward medical care, to their beliefs about the cause of disease and the ability to alter it through public health and medical interventions, this country is profoundly different from the United States. You ask: how does culture affect health?

You are working in a country with strict Islamic practices and find that religion, like culture, can have major impacts on health. Religious practices differ widely—from beliefs about food and alcohol; to sexual practices, such as male circumcision and female sexual behavior; to acceptance or rejection of interventions aimed at women's health. You ask: how does religion affect health?

You're trying to help your spouse quit smoking cigarettes and your kids from starting. You know that gentle encouragement and support on a one-on-one basis are essential, but often are not enough because cigarettes are an addiction that produces withdrawal and long-term cravings. Like most addictions, it requires a combination of individual motivation, support from family and friends, and sometimes the use of medications; but you wonder: do warning labels on cigarettes, taxes on cigarettes, and "no smoking" zones in public places make any difference?

Your efforts to convince your friends to avoid smoking (or at least stop smoking) focus on giving them the facts about how cigarettes cause lung cancer, throat cancer, and serious heart disease. You are frustrated by how little impact you have on your friends. You wonder, whether you would be more successful by focusing on the immediate negative impacts, such as stained teeth and bad breath, as well as the loss of control that goes along with addiction to nicotine?

As a new parent, you hear from your pediatrician, nurses in the hospital, and even from the makers of your brand of diapers, that babies should sleep on their backs. They call it "back-to-sleep." You're surprised to find that it's part of the class on babysitting given by the local community center and a required part of the training for those who work in registered day care centers. You find out that it's all part of a social marketing campaign that has reduced in half the number of deaths from sudden infant death syndrome. You ask: why has the "back-to-sleep" campaign been so successful?

Each of these cases illustrates ways that an understanding of social and behavioral sciences can contribute to an understanding of public health. Let us explore these connections.

HOW IS PUBLIC HEALTH RELATED TO THE SOCIAL AND BEHAVIORAL SCIENCES?

The development of social and behavioral sciences in the 19th and 20th centuries is closely connected with the development of public health. These subject areas share a fundamental belief that understanding the organization and motivation behind social forces, along with a better understanding of the behavior of individuals, can be used to improve the lives of individuals, as well as those of society as a whole.[1]

The 19th century development of social and behavioral sciences, as well as public health, grew out of the Industrial

TABLE 6-1 Examples of Contributions of Social and Behavioral Sciences to Public Health

Social science discipline[a]	Examples of disciplinary contributions to public health
Psychology	Theories of the origins of behavior and risk taking tendencies and methods for altering individual and social behaviors
Sociology	Theories of social development, organizational behavior, and systems thinking. Social impacts on individual and group behaviors
Anthropology	Social and cultural influences on individual and population decision making for health with a global perspective
Political science/Public policy	Approaches to government and policy making related to public health. Structures for policy analysis and the impact of government on public health decision making
Economics	Understanding the micro- and macroeconomic impact on public health and health care systems
Communications	Theory and practice of mass and personalized communication and the role of media in communicating health information and health risks
Demography	Understanding demographic changes in populations globally due to aging, migration, and differences in birth rates, plus their impact on health and society
Geography	Understanding of the impacts of geography on disease and determinants of disease, as well as methods for displaying and tracking the location of disease occurrence

[a] A similar list of contributions of the humanities could be developed including the contributions of literature, the arts, history and philosophy, and ethics. These contributions of the social sciences are in addition to contributions of the sciences, mathematics, and humanities. Biology, chemistry, and statistics underpin much of epidemiology and environmental health. Languages and culture, history, and the arts provide key contributions to health communications and health policy. Thus, a broad arts and sciences education is often considered a key part of preparation for public health.

Revolution in Europe, and later in America. It was grounded in efforts to address the social and economic inequalities that developed during this period and provided an intellectual and institutional structure for what was and is now called social justice. Social justice implies a society that provides fair treatment and a fair share of the rewards of society to individuals and groups of individuals. Early public health reformers advocated for social justice and saw public health as an integral aspect of it.

The intellectual link between social and behavioral sciences and public health is so basic and so deep that it is often taken for granted. As students with opportunities to learn about both social sciences and public health, it is important to understand the key contributions that social sciences make to public health. It is not an exaggeration to view public health as an application of the social sciences, i.e., as an applied social science. Table 4–1 summarizes many of the contributions that the social sciences make to public health.

Let us start with a key question that needs to be addressed to understand the relationship between social forces and health.

How Does Socioeconomic Status Affect Health?

Beginning in the 19th century, social scientists developed the concept of **socioeconomic status**. They also developed elaborate systems to operationalize the definition of socioeconomic status and classify individuals. In the United States, the definition has generally included measures that are primarily economic including:[b]

- family income
- educational level or parents' educational level
- professional status or parents' professional status

Let us examine how socioeconomic status has been shown to affect health. Then, we will examine additional social factors that affect health and the response to disease.

Health status, at least as measured by life expectancy, is strongly associated with socioeconomic status.[2, 3] Greater longevity is associated with higher social status with a gradient of increasing longevity from lowest to highest on the socioeconomic scale.

It is also important to recognize that socioeconomic impact are not solely related to a person's income Above an annual threshold income level of about $10,000 per person, the association of longevity with income is best explained by the disparities in income, rather than the absolute level. Thus, developed countries with smaller disparities of income, such as Japan, Sweden, and Canada, have greater average longevity and smaller disparities in longevity between their richest and poor-

[b] A more formal social hierarchy has traditionally existed in Europe. European social scientists utilized the concept of social class when categorizing individuals by socioeconomic status. In Europe, economics alone was not thought to be adequate to explain socioeconomic status or categorize individuals.

est citizens than compared to a country like the United States. In the United States, greater disparities in income and longevity exist between the richest and poorest citizens. However, the enormous diversity of the population of the United States in terms culture and religion as well as socioeconomic level may also help explain the disparities in longevity.[c]

We understand many, but not all, of the ways that socioeconomic factors affect health. Greater economic wealth usually implies access to healthier living conditions. Improved sanitation, less crowding, greater access to health care, and safer methods for cooking and eating are all strongly associated with higher economic status in developed, as well as developing countries.

Education is also strongly associated with better health. It may change health outcomes and increase longevity by encouraging behaviors that provide protection against disease and likewise reduce exposure to behaviors that put individuals at risk of disease. Higher education levels, coupled with the increased resources that greater wealth can provide, may increase access to better medical care and provide greater ability to protect against health hazards.

Individuals of lower socioeconomic status are more likely to be exposed to health hazards at work and in the physical environment through toxic exposure in the air they breathe, in the water they drink, and in the food they eat. Table 6-2 outlines a number of mechanisms by which socioeconomic status can directly and indirectly influence health.

These factors, while important, explain only about half of the observed differences in life expectancy among individuals of different socioeconomic status. For instance, the rates of coronary heart disease are considerably higher among those of lower socioeconomic status—even after taking into account cigarette smoking, high blood pressure, cholesterol levels, and blood sugar counts.[3]

Considerable research is now being directed to better understand these and other effects of socioeconomic status. One theory suggests that social control and social participation may help explain these substantial differences in health. It contends

that control over individual and group decision making is much greater among individuals of higher socioeconomic status. The theory holds that the ability to control one's life may be associated with biological changes that affect health and disease.[3] Additional research is needed to confirm or reject this theory and/or provide an adequate explanation for these important, yet unexplained, differences in health based upon socioeconomic status.

What Other Social Factors Explain Differences in Health and Response to Disease?

Culture and religion have effects on health above and beyond socioeconomic status as measured by income, education, and professional status. They can be viewed as behaviors, values, and beliefs that are learned from others and shared with others.

Culture

Culture, in a broad sense, helps people make judgments about the world and decisions about behavior. Culture defines what is good or bad, and what is healthy and unhealthy. This may relate to lifestyle patterns, beliefs about risk, and beliefs about body type—for example, a large body type in some cultures symbolizes health and well-being, not overweight or other negative conditions.

Culture directly affects the daily habits of life. Food choice and methods of food preparation and preservation are all affected by culture, as well as socioeconomic status. The Mediterranean diet, which includes olive oil, seafood, vegetables, nuts, and fruits, has been shown to have benefits for the heart even when used in countries far removed from the Mediterranean.

There are often clear-cut negative and/or positive impacts on disability related to cultural traditions as diverse as feet binding in China and female genital mutilation in some parts of Africa. Some societies reject strenuous physical activity for those who have the status and wealth to be served by others.

Culture is also related to an individual's response to symptoms and acceptance of interventions. In many cultures, medical care is exclusively for those with symptoms and is not part of prevention. Many traditional cultures have developed sophisticated systems of self-care and self-medication supported by family and traditional healers. These traditions greatly affect how an individual responds to symptoms, how they communicate the symptoms, and the types of medical and public health interventions that they will accept.

Many cultures allow and even encourage the use of traditional approaches alongside Western medical and public health approaches. In some cultures, traditional healers are considered appropriate for health problems whose causes are not thought to be biological, but related to spiritual and other phenomena. Recent studies of alternative, or complementary, medicine have provided evidence that specific traditional

---

[c] The association between socioeconomic status and longevity is most strongly associated with an individual's socioeconomic status as an adult. The socioeconomic status of an individual's parents has a much weaker association. This suggests that genetic factors have little to do with the association between socioeconomic status and life expectancy. Education has a stronger association with health status than income or professional status. Lower socioeconomic status leads to poor health rather than poor health leading to lower socioeconomic status. Socioeconomic factors are associated with an increase in relative risk of death of 1.5 to 2.0 when comparing the lowest and highest socioeconomic groups. This means that those in the lowest group have more than a 50 percent increase in the death rate compared to the highest group. This relative risk steadily increases as the socioeconomic level decreases. The relationship has a dose-response relationship, that is, there is an increase in longevity with every increase in socioeconomic status. Thus, the impact is not limited to those with the lowest status. The largest contributors to the differences in the death rate are: cardiovascular disease, violence, and increasingly AIDS; however the death rate is impacted in general by a wide range of diseases—most being malignancies and infectious diseases.[3]

TABLE 6-2 Examples of Ways that Socioeconomic Status May Affect Health

Type	Examples
Living conditions	Increases in sanitation, reductions in crowding, methods of heating and cooking
Overall educational opportunities	Education is the strongest association with health behaviors and health outcomes.
	May be due to better appreciation of factors associated with disease and greater ability to control these factors
Educational opportunities for women	Education for women has an impact on the health of children and families
Occupational exposures	Lower socioeconomic jobs are traditionally associated with increased exposures to health risks
Access to goods and services	Ability to access goods, such as protective devices and high quality foods and services, including medical and social services to protect and promote health
Family size	Large family size affects health and is traditionally associated with lower socioeconomic status and with lower health status
Exposures to high risk behaviors	Social alienation related to poverty may be associated with violence, drugs, other high risk behaviors
Environmental	Lower socioeconomic status associated with greater exposure to environmental pollution, "natural" disasters, and dangers of the "built environment"

interventions, such as acupuncture and specific osteopathic and chiropractic manipulation, have measurable benefits. Thus, cultural differences should not be viewed as problems to be addressed, but rather as practices to be understood. Table 6-3 summarizes a number of the ways that culture can affect health.

Religion

Social factors affecting health include religion along with culture. Religion can have a major impact on health particularly for specific practices that are encouraged or condemned by a particular religious group. For instance, we now know that male circumcision reduces susceptibility to HIV/AIDS. Religious attitudes that condone or condemn the use of condoms, alcohol, and tobacco have direct and indirect impacts on health as well.

Some religions prohibit specific healing practices, such as blood transfusions or abortion, or totally reject medical interventions altogether, as is practiced by Christian Scientists. Religious individuals may see medical and public health interventions as complementary to religious practice or may substitute prayer for medical interventions in response to symptoms of disease. Table 6-4 outlines some of the ways that religion may affect health.

We have examined a number of ways that socioeconomic, cultural, and religious factors may affect health and the response to disease. Many, but not all, of these factors ultimately influence the health-related behavior of individuals. Thus, we need to look at the relationship between behavior and health and the ways that we can use knowledge from the social sciences to improve health.

Can Health Behavior Be Changed?

Much of the preventable disease and disability today in the United States and other developed countries is related to the behavior of individuals. From cigarette smoking to obesity, from intentional to unintentional injuries, from sexual behavior to drug abuse, health issues can be traced to the behavior of individuals. At times, we hear discouraging messages that behavior cannot be changed. However, if we take a relatively long-term view, we find that there are many examples of behavioral change that have occurred for the better. For instance:

- Cigarette smoking in the United States among males has been reduced from approximately 50 percent in the 1960s to less than 25 percent today.
- Infants today generally are placed on their backs for sleeping and napping and not on their stomachs, as was the usual practice in the 1980s and earlier. "Back-to-sleep" campaigns are believed to have reduced Sudden Infant Death Syndrome (SIDS) by nearly 50 percent in the United States.
- Seat belt use in the United States has increased from less than 25 percent in the 1970s to over 80 percent currently.
- Drunk driving in the United States has been dramatically reduced with a resulting decline in automobile-related fatalities.
- Mammography use increased by approximately 50 percent during the 1990s and has been credited with beginning to reduce the previously-rising mortality rates from breast cancer.

TABLE 6-3 Examples of Ways that Culture Can Affect Health

Ways that culture may affect health	Examples
Culture is related to behavior—social practices may put individuals and groups at increased or reduced risk	Food preferences—vegetarian, Mediterranean diet Cooking methods History of binding of feet in China Female genital mutilation Role of exercise
Culture is related to response to symptoms, such as the level of urgency to recognize symptoms, seek care, and communicate symptoms	Cultural differences in seeking care and self-medication Social, family, and work structures provide varying degree of social support—low degree of social support may be associated with reduced health-related quality of life
Culture is related to the types of interventions that are acceptable	Variations in degree of acceptance of traditional Western medicine including reliance on self-help and traditional healers
Culture is related to the response to disease and to interventions	Cultural differences in follow-up, adherence to treatment, and acceptance of adverse outcome

The potential to change behavior can make health worse as well. The following changes for the worse have also occurred in the United States in recent years:

- Over the last three decades, Americans have increased their caloric intake and reduced their average amount of exercise, resulting in a doubling of the obesity rate to approximately one-third of all adults.
- Between the 1960s and the 1990s, teenage girls and young adult women increased their cigarette smoking, subjecting their unborn children to additional hazards of low birthweight.

TABLE 6-4 Examples of Ways that Religion May Affect Health

Ways that religion affects health	Examples
Religion may affect social practices that put individuals at increased or reduced risk	Sexual: circumcision, use of contraceptive Food: avoidance of seafood, pork, beef Alcohol use: part of religion versus prohibited Tobacco use: actively discouraged by Mormons and Seventh-Day Adventists as part of their religion
Religion may affect response to symptoms	Christian Scientists reject medical care as a response to symptoms
Religion may affect the types of interventions that are acceptable	Prohibition against blood transfusions Attitudes toward stem cell research Attitudes toward abortion End-of-life treatments
Religion may affect the response to disease and to interventions	Role of prayer as an intervention to alter outcome.

Thus, behaviorial change is possible for the better and for the worse. Some behaviors, however, are easier to change than others. Let's take a look at why this is.

Why Are Some Individual Health Behaviors Easier to Change Than Others?

Some behaviors are relatively easy to change, while others are extremely difficult. Being able to recognize the difference is an important place to start when trying to alter behavior. It is relatively easy when one behavior can be substituted for a similar one and results in a potentially large payoff. In these situations, knowledge often goes a long way. For instance, the substitution of acetaminophen (Tylenol) for aspirin to prevent Reye's Syndrome was relatively easy. Similarly, the "back-to-sleep" campaign was quite successful in reducing the rate of death from SIDS. In both of these cases, an acceptable and convenient substitute was available making the needed behavioral change much easier to accomplish.

Along with knowledge, incentives—such as reduced cost, increased availability, or improvements in ease-of-use—can encourage rapid acceptance and motivate behavioral change. For instance, easier-to-install child restraint systems have increased their use. Greater insurance coverage and widespread availability of modern mammography equipment has led to an increase in the number of mammograms performed.

The most difficult behaviors to change are those that have a physiological component, such as obesity, or an addictive element, such as cigarette smoking. Individual interventions aimed at smoking cessation or long-term weight control generally succeed less than 30 percent of the time—even among motivated individuals. Even intensive interventions with highly-motivated individuals cannot be expected to be successful more than 50 percent of the time, as was illustrated by the Multiple Risk Factor Intervention Trial (MRFIT), which attempted intensive interventions to reduce risk factors for cardiovascular disease.

In addition, physical, social and economic barriers can stand in the way of behavior change, even if individuals themselves are motivated. If health care is not accessible, or if survival needs require individuals to engage in risks they might not take otherwise, change in behavior may be impeded.

Successful behavioral change requires that we understand as much as we can about how behavior can be changed and what we can do to help.

How Can Individual Behavior Be Changed?

The behavior of individuals is often the final common pathway through which disease, disability, and death can be prevented. The fact that individual behavior has a clearly observable connection with these factors does not necessarily imply that the best or only way to address the behavior of individuals is to focus exclusively on individuals. The forces at work to mold individual behaviors are sometimes referred to as **downstream factors**, **mainstream factors**, and **upstream factors**. Downstream factors are those that directly involve an individual and can potentially be altered by individual interventions, such as an addiction to nicotine. Mainstream factors are those that result from the relationship of an individual with a larger group or population, such as peer pressure to smoke or the level of taxation on cigarettes. These factors require attention at the group or population level. Finally, upstream factors are often grounded in social structures and policies, such as government-sponsored programs that encourage tobacco production. These require us to look beyond traditional health care and public health interventions to the broader social and economic forces that affect health.

Thus, changes in behavior often require more than individual motivation and determination to change. They require encouragement and support from groups ranging from friends and families to work and peer groups. Behavioral change may also require social policies and expectations that reinforce individual efforts. It also requires us to examine the stages that individuals experience as they struggle to change behaviors, especially those habits with physiological challenges.

What Stages Do Individuals Go Through in Making Behavioral Changes?

The process of individual behavioral changes can be described using what has been called the **Stages of Change model**.[4–6] This model provides a useful longitudinal description of five steps that individuals go through in changing behavior. It also suggests steps that can be taken to help individuals make changes.[d]

The first stage, called **precontemplation**, implies that an individual has not yet considered changing their behavior. At this stage, efforts to encourage change are not likely to be successful. However, efforts to educate and offer help in the future may lay the groundwork for later success.

The second phase, known as **contemplation**, implies that an individual is actively thinking about the benefits and barriers to change. At this stage, information focused on short- and immediate-term gains, as well as long-term benefits, can be especially useful. In addition, the contemplation stage lends itself to developing a baseline—that is, establishing the current severity or extent of the problem in order to measure future progress.

---

[d]Designing interventions based upon the Stages of Change model has not been uniformly successful. One potential reason for this may be that individuals are often in different stages of change for different types of interventions. With complex interventions, such as those required to address obesity, an individual may be in one stage of change for exercise, another for adding fruits and vegetables to his/her diet, and in a different stage in terms of calorie reduction.

The third phase is called **preparation**. During this phase the individual is developing a plan of action. At this point, the individual may be especially receptive to setting goals, considering a range of strategies, and developing a timetable. Help in recognizing and preparing for unanticipated barriers can be especially useful to the individual during this phase.

The fourth phase is the **action** phase when the change in behavior takes place. This is the time to bring together all possible outside support to reinforce and reward the new behavior and help with problems or setbacks that occur.

The fifth—and hopefully final phase—is the **maintenance** phase in which the new behavior becomes a permanent part of an individual's lifestyle. The maintenance phase requires education on how to anticipate the long-term nature of behavioral change, especially how to resist the inevitable temptations to resume the old behavior. Using the cigarette smoking as illustration again, Table 4–5 summarizes the stages of behavioral change and the specific actions that can be helpful at each of the stages.

Notice that individual behavioral change is made easier when group and social efforts are also brought to bear. Higher taxes on cigarettes, peer attitudes and support, and enforcement of laws restricting the permissible locations for smoking, all can be effective social interventions that increase the chances of entering the precontemplation phase, reinforce individual efforts to change behavior, and assist with the maintenance phase. Thus, when looking at how to change individual behavior, we also need to consider how group behaviors can be changed.

How Can Group Behaviors Be Changed?

In recent years, public health has begun to apply marketing approaches to try to better understand and change the health behaviors of groups of people—especially those like cigarette smokers who are at high risk of health impacts of their behavior. **Social marketing**, a use and extension of traditional product marketing, has become a key component of a public health approach to behavioral change.[7] Social marketing campaigns were first successfully used in the developing world for promoting a range of products and behaviors, including family planning and pediatric rehydration therapy. In recent years, social marketing efforts have been widely and successfully used in developed countries, including such efforts as:

- The truth® campaign—Developed by the American Legacy Foundation, it aims to redirect smoking from being seen as a teenage rebellion to not smoking being a rebellion against the alleged behavior-controlling tobacco industry.
- The National Youth Anti-Drug campaign—It uses social marketing efforts directed at young people, including the "Parents. The anti-drug." campaign.

- The VERB™ campaign—It focused on 9 to 13 year olds, or "tweens," with a goal of making exercise fun and "cool" for everyone, not just competitive athletes.

Social marketing incorporates the "4 Ps," which are widely used to structure traditional marketing efforts. These are:

- **Product**: Identifying the behavior or innovation that is being marketed
- **Price**: Identifying the benefits, the barriers, as well as the financial costs
- **Place**: Identifying the target audiences and how to reach them
- **Promotion**: Organizing a campaign or program to reach the target audience(s)

Social marketing has incorporated concepts from the **diffusion of innovation theory**. This theory, like the stages of behavioral change, contends that adoption of new behaviors requires a series of phases or steps. These move from knowledge of the innovation, to persuasion of its benefits, to the decision to adapt, to implementation, and confirmation.[1] The diffusion of innovation theory has contributed the concept of different types of adopters including: **early adopters**—those who seek to experiment with innovative ideas; **early majority adopters**—often opinion leaders whose social status frequently influences others to adopt the behavior; and **late adopters** (or laggards)—those who need support and encouragement to make adoption as easy as possible.

A different approach is often needed to engage each of these groups. For instance, marketing efforts may initially target early adopters with an approach encouraging innovation and creativity. This may be followed by an approach to opinion leaders who can help the innovation or behavior change become mainstream. A different approach emphasizing ease-of-use and widespread acceptance may be most helpful for encouraging late adopters.

Social marketing, like product marketing, often relies on what marketers call **branding**. Branding includes words and symbols that help the target audience identify with the service; however, it goes deeper than just words and symbols. It can be seen as a method of implementing the fourth "P," or promotion. It also builds upon the first three "Ps":

- Branding requires a clear understanding of the product or the behavior to be changed (product).
- Successful branding puts forth strategies for reducing the financial and psychological costs (price).
- Branding identifies the audience and segments of the audience and asks how each segment can be reached (place).

Thus, branding is the public face of social marketing, but it also needs to be integrated into the core of the marketing plan.[e]

TABLE 6-5 Stages of Behavioral Change

Stages of change	Actions	Example—Cigarette Smoking
Precontemplation	**Prognosticate**	
Individuals not considering change	Assessing readiness for change—timing is key	Determine individual's readiness to quit. If not ready, indicate receptivity to help in the future
		Look for receptive timing such as during acute respiratory symptoms
		Social factors, such as workplace and indoor restriction on smoking and taxation, increase likelihood of entering precontemplation phase
Contemplation	**Motivate change**	
Individual thinks actively about the health risk and action required to reduce that risk	Provide information focused on short and intermediate gains from behavioral change, as well as long-term benefits	Reinforce increase in exercise level, reduction in cough, financial savings, serving as example to children, protection of fetus, etc.
Issue of change is on the individual's agenda but no action planned	Doubtful, dire, and distant impacts are less effective	Also continue to inform of longer term effects on health
	Establish baseline to assess severity of the problem; focus attention on the problem and provide basis for comparison	Develop log of timing, frequency, and quantity of smoking, as well as associated events
Preparation	**Plan change**	
Prepare for action including developing a plan and setting a timetable	Set specific measurable and obtainable goals with deadlines	Quit date or possible tapering if heavy smoker
	Two or more well chosen simultaneous interventions may maximize effectiveness	Family support, peer support, individual planning, medication, etc.—may reinforce and multiply impacts
	Recognize habitual nature of existing behavior and remove associated activities	Remove cigarettes, ashtrays, and other associated smoking equipment
		Remove personal and environmental impacts of past smoking, such as teeth cleaning and cleaning of drapery
		Anticipate temptations, such as associations with food, drink, and social occasions
Action	**Reinforce change**	
Observable changes in behavior with potential for relapse	Provide/suggest tangible rewards	Provide rewards, such as alternative use of money—focus on personal hygiene or personal environment
	Positive feedback encouragement of new behavior	Focus on measurable progress toward new behavior
	Anticipate adverse effects and frustrations	Provide receptive environment, but avoid focus on excuses
		Take short term one-day-at-a-time approach
		Recognize cravings and have plan including use of medications
		Recognize potential for symptoms to worsen at first before improvement occurs
		Anticipate potential for weight gain and encourage exercise and other behaviors to reduce potential for weight gain
	Utilize group/peer support	Family and peer reinforcement critical during action phase

continues

TABLE 6-5 Stages of Behavioral Change (continued)

Maintenance	Maintain change	
New behavior needs to be consolidated as part of permanent lifestyle change	Practice/reinforce methods for maintaining new behavior	Avoid old associations and prepare/practice response when encountering old circumstances
	Recognize long term nature of behavioral change and need for supportive peers and social reinforcement	Negative social attitudes toward smoking among peers and society along with social restrictions, such as limiting public indoor smoking and social actions, such as taxation, help prevent smoking and reinforce maintenance of cessation

Source: Data from Prochaska JO, DiClemente CC. Stages and processes of self-change of smoking: toward an integrative model of change. *J Consult Clin Psychol.* 1983;51:390–395.

Social marketing efforts in developing and developed countries have demonstrated that it is possible to change key health behaviors of well-defined groups of people, including adolescents, which are often regarded as the hardest to reach. An example of the use of social marketing to reach young people, the VERB™ campaign, is examined in Box 6-1.[8]

To be most successful, behavior-changing efforts need to include individual, group, and social aspects. Let us look at some approaches to combining these efforts.

How Can We Combine Individual, Group, and Social Efforts to Implement Behavioral Change?

As we have seen, behavioral change is most successful when it combines efforts aimed at the individuals, the at-risk group, and the population, or society as a whole. It can be useful to look again at the Stages of Change model to see how interventions can be successfully combined at each stage in the behavior change process. Let us return to cigarette smoking as an example.

---

[e] Social marketing has not only incorporated traditional product marketing approaches, it has extended them to address the special circumstances of not-for-profit and government organizations. The use of social marketing in public health has required modifications and enhancements that have been described using four more "Ps": publics, partnerships, policies, and purse strings. "Publics" refers to the need to reach not only a target audience whose behavior we seek to change, but also those people who influence the target audience—be they parents, employers, or opinion leaders. For example, a campaign to address obesity, cigarettes, or high-risk sexual behavior in schools, requires support from parents. "Partnerships" refers to the need for collaborations to achieve most public health goals. The VERB™ campaign, for instance, partnered with television stations appealing to "tweens" and schools to help get its message out. Successful efforts to reduce adolescent smoking, increase exercise, and reduce drug use, also require changing institutional policies, which means reaching adult decision makers. Finally, the "purse string" aspect is money—few public health social marketing campaigns have adequate resources to do the job. Funding issues may require public health marketing teams to incorporate a long-term approach and look for nontraditional sources of funding.

BOX 6-1 VERB™ Campaign.

The VERB™ social marketing campaign was funded through the Centers for Disease Control and Prevention (CDC), which worked with advertising agencies to reach "tweens" to make exercise "cool." After a series of focus groups and other efforts to define and understand the market, they concluded that the message should not be one of improving health, but rather of having fun with friends, exploring new activities with a sense of adventure, and being free to experiment without being judged on performance.

Marketing efforts also identified barriers including time constraints and the attraction of other activities from social occasions to television to computers. Barriers included lack of access to facilities, as well as negative images of competition, embarrassment, and the inability to become an elite athlete.

The VERB™ campaign implied action and used the tagline *"It's what you do."* Initial messages used animated figures of children covered with verbs being physically active. Later, messages turned these animated verb-covered kids into real kids actively playing. Widely-used logos were developed and promoted as part of the branding effort. The VERB™ campaign partnered with television channels that successfully reach "tweens," sponsored outreach events, and distributed promotional materials.

During the four years of the VERB™ campaign, "tweens" developed widespread recognition of the program and rated it highly in terms of "saying something important to me" and "makes me want to get more active" with maximum levels of recognition of 64 percent and 68 percent, respectively. Despite the documented success of VERB™, it was discontinued because of cuts in the federal budget.

TABLE 6-6 Stages of Change—Individual, Group and Population/Social Interventions to Change Cigarette Smoking Behavior

Stage of change	Individual	At-risk group	Population/society
Precontemplation	Assess readiness for change and offer future help	Social marketing aimed at specific groups Restriction on smoking at work	Cost affected by taxes, restrictions on smoking in public places, warning labels on packages
Contemplation	Information on hazards of smoking and gains from quitting	More receptive to social marketing aimed at specific groups Restriction on smoking at work	More receptive to costs of cigarettes, restrictions on smoking in public places, and warning labels
Preparation	Set individual goals and develop strategy Medication may be helpful	Support group/friends and family reinforce individual preparation; telephone "quit lines"	National efforts, e.g., American Cancer Society National Quit Day
Action	Remove connections between cigarettes and pleasurable activities Use of medications if needed	Public commitment to action—announce to family, friends, and work colleagues	Pay for medication and other assistance with cessation as part of insurance
Maintenance	Education regarding long term physical addiction and potential for relapse	Continued reinforcement at work and by peer and social groups	Continued reinforcement by social marketing, taxes, and restriction on public smoking

In the precontemplation and contemplation stages, individual interventions are focused on education, assessing readiness to change, and offering help. Interventions targeting at-risk groups and populations, such as taxation on cigarettes and restriction on public and workplace smoking, can be very useful in smoking cessation preparation and preventing individuals from starting smoking in the first place.

The preparation stage requires individual action, but it can be encouraged and reinforced by family and friends, as well as via national efforts, such as the American Cancer Society's annual Great American Smokeout®, which encourages smokers to quit for a lifetime by starting with just one day. The action phase may appear to be exclusively based on individual action; however, it can be supported and encouraged by family and peers and reinforced by social efforts, such as health insurance polices that provide payment for support groups and medications.

The maintenance phase also relies on individual, group, and population/social interventions. Individual interventions often focus on education about the long-term nature of behavioral change and the efforts necessary to resist the temptation to resume smoking. Group support and reinforcement continue to be important in encouraging maintenance of the new behavior. The same types of social interventions that encourage individuals to stop cigarette smoking, such as taxation and restrictions on public and workplace smoking, also help encourage them to continue their smoke-free behavior. Table 6-6 summarizes the roles that individual, group, and population intervention can play in changing cigarette smoking behavior.

We have now looked at how health informatics and health communications, as well as social and behavioral interventions, can be used in population health.

Key Words

- Stages of Change model
- Social marketing
- Diffusion of innovation theory
- Branding
- Downstream factors
- Mainstream factors
- Upstream factors
- Precontemplation
- Contemplation
- Preparation
- Action
- Maintenance
- Product
- Price
- Place
- Promotion
- Early adopters
- Early majority adopters
- Late adopters

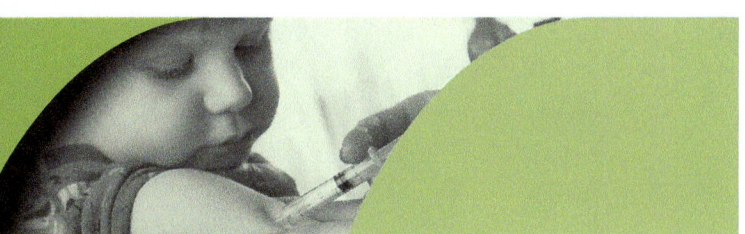

Discussion Questions

Take a look at the questions posed in the following scenarios, which were presented at the beginning of this chapter. See now whether you can answer these questions.

1. You travel to a country in Asia and find that their culture affects most parts of life. From the food they eat and their method of cooking to their attitudes toward medical care to their beliefs about the cause of disease and the ability to alter it through public health and medical interventions, this country is profoundly different from the United States. You ask: how does culture affect health?

2. You are working in a country with strict Islamic practices and find that religion like culture can have major impacts on health. Religious practices differ widely—from beliefs about food and alcohol; to sexual practices, such as male circumcision and female sexual behavior; to acceptance or rejection of interventions aimed at women's health. You ask: how does religion affect health?

3. You're trying to help your spouse quit smoking cigarettes and your kids from starting. You know that gentle encouragement and support on a one-on-one basis are essential, but often are not enough because cigarettes are an addiction that produces withdrawal and long-term cravings. Like most addictions, it requires a combination of individual motivation, support from family and friends, and sometimes the use of medications; but you wonder: do warning labels on cigarettes, taxes on cigarettes, and "no smoking" zones in public places make any difference?

4. Your efforts to convince your friends to avoid smoking (or at least stop smoking) focus on giving them the facts about how cigarettes cause lung cancer, throat cancer, and serious heart disease. You are frus-

trated by how little impact you have on your friends. You wonder, whether you would be more successful by focusing on the immediate negative impacts, such as stained teeth and bad breath, as well as the loss of control that goes along with addiction to nicotine?

5. As a new parent, you hear from your pediatrician, nurses in the hospital, and even from the makers of your brand of diapers, that babies should sleep on their backs. They call it "back-to-sleep." You're surprised to find that it's part of the class on babysitting given by the local community center and a required part of the training for those who work in registered day care centers. You find out that it's all part of a social marketing campaign that has reduced in half the number of deaths from sudden infant death syndrome. You ask: why has the "back-to-sleep" campaign been so successful?

REFERENCES

1. Edberg M. *Essentials of Health Behavior: Social and Behavioral Theory in Public Health.* Sudbury, MA: Jones and Bartlett Publishers; 2007.
2. Commission on Social Determinants of Health. *Closing the Gap in a Generation: Health Equity through Action on the Social Determinants of Health. Final Report of the Commission on Social Determinants of Health.* Geneva, Switzerland: World Health Organization; 2008.
3. Marmot M. *The Status Syndrome: How Social Standing Affects Our Health and Longevity.* New York: Henry Holt and Company; 2004.
4. DiClemente CC, Prochaska JO. Self-change and therapy change of smoking behavior: a comparison of processes of change in cessation and maintenance. *Addict Behav.* 1982;7:133–142.
5. Prochaska JO, DiClemente CC. Stages and processes of self-change of smoking: toward an integrative model of change. *J Consult Clin Psychol.* 1983;51:390–395.
6. Jenkins CD. *Building Better Health: A Handbook of Behavioral Change.* (Technical Publication No. 590). Washington, DC: Pan American Health Organization; 2003.
7. Weinreich NK. What is social marketing? Available at: http://www.social-marketing.com/Whatis.html. Accessed March 12, 2009.
8. Wong F, Huhman M, Asbury L, Mueller RB, McCarthy S, Londe P, et al. VERB™—A social marketing campaign to increase physical activity among youth. *Prev Chronic Disease.* 2004;3(1):1–7.

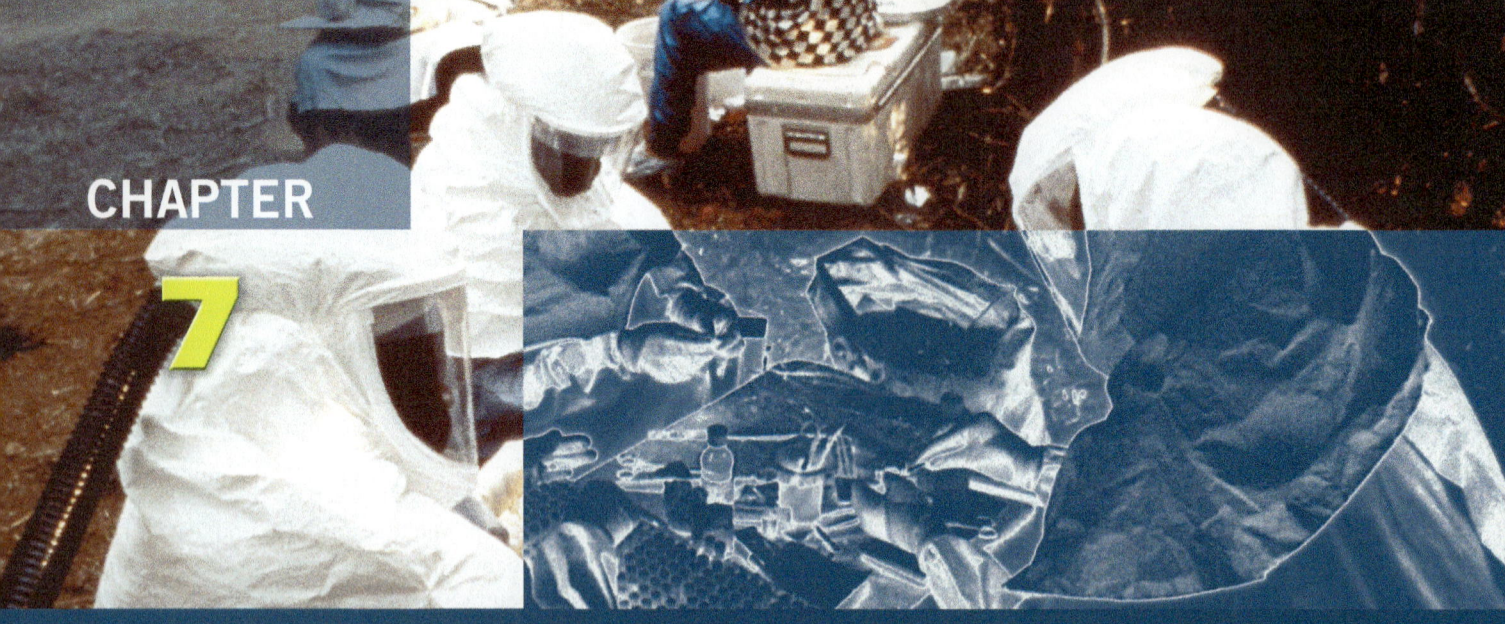

Identifying the Challenge

If human civilization lasts, if it continues to spread, infectious diseases will increase in number in every region of the globe. Exchanges and migrations will bring the human and animal diseases of every country. The work is already well advanced; its future is assured.

—Charles Nicolle, *The Destiny of Infectious Diseases, 1932*

Preview

This book is about a challenge—a worldwide challenge—posed by microbes, invisible marauders that inhabit the earth, perhaps causing illness and death. Classically, there are five distinct kinds of microbes: **bacteria**, **viruses**, **protozoans**, **fungi**, and **unicellular algae**. **Prions** can now be added to the list, bringing the total to six, perhaps better called "infectious agents." It should be emphasized at the outset that a relatively few members of each of the groups pose the potential for infection. Most microbes are often beneficial and are essential to the cycles of nature without which higher life forms could not exist. In many cases, microbes have been harnessed for the benefit of humankind.

But this book, by intent, has a bias because its theme relates to those few microbes that are disease producers. In the language of medical microbiology, they are referred to as **pathogens** or virulent microbes. Why some of these microbial diseases now represent an increased challenge is the subject matter of this chapter.

96

The Challenge

Forty or so years ago, there was less need for this book than there is today. However, in the decades since, forty previously unknown infectious diseases have emerged and others have reemerged: AIDS, Ebola virus, *Escherichia coli*, hantavirus, West Nile virus, *Salmonella*, flesh-eating "strep," and "mad cow disease," to name only a few (FIGURE 7.1). Movies, books, and articles about microbial diseases intrigue large numbers of viewers and readers. Popular news magazine programs, including "Dateline," "60 Minutes," and "20/20," frequently air segments relating to dangerous microbes; newspaper articles and news broadcasts appear almost daily and further alert the public to threats posed by microbes.

The 1990s were especially eventful. In 1992 tuberculosis (TB), a bacterial disease almost relegated to oblivion, reemerged in New York City, resulting in almost four thousand cases; the tubercle bacillus was developing resistance to a variety of antibiotics that had once stopped the bacteria dead in their tracks. In 1993 an outbreak of cryptosporidiosis, a waterborne protozoan disease characterized by diarrhea, swept through Milwaukee, Wisconsin, causing illness in about 400 thousand people, approximately 25% of that city's population; it was the largest reported waterborne illness in U.S. history. In 1993 hantavirus, the causal agent of a potentially lethal influenza-like respiratory illness, reemerged with deadly results in New Mexico, Utah, Colorado, and Arizona. Ebola hemorrhagic fever, caused by one of the deadliest known viruses, ignited a panic in 1995 when 240 people bled to death during an outbreak that occurred in Kikwit, Zaire (now the Democratic Republic of Congo). Earlier, in the winter of 1989, scientists working at a primate quarantine facility just outside Washington, DC were terrified when

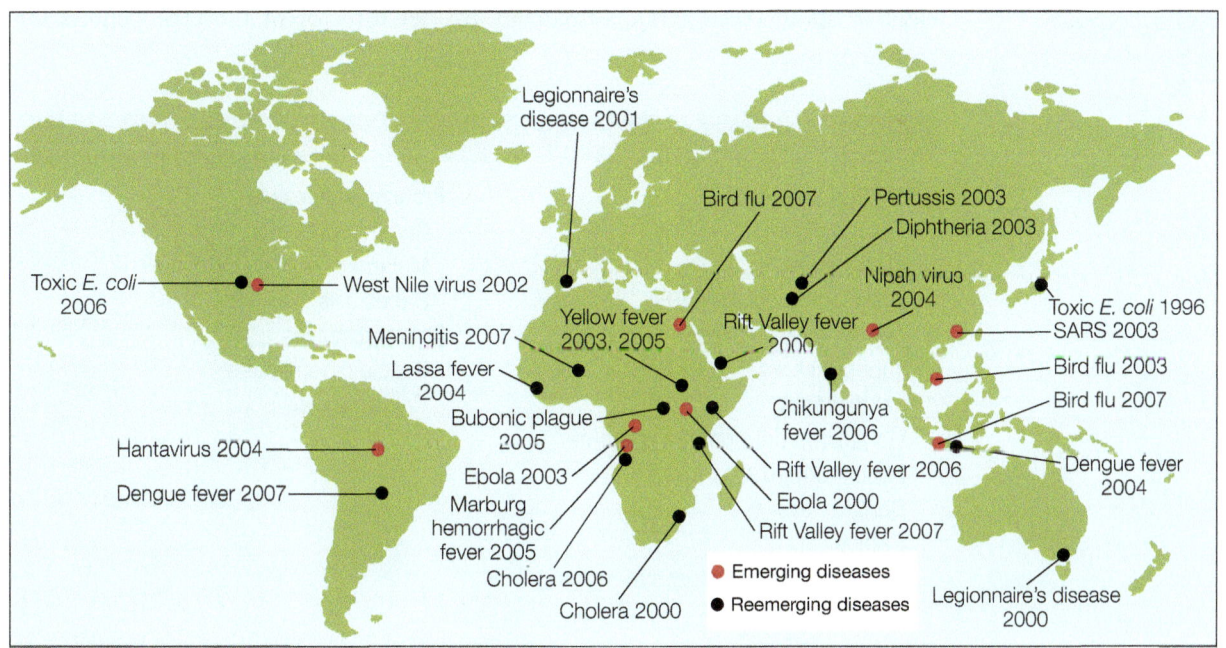

FIGURE 7.1   Emerging and reemerging diseases, 1996–2007.

Ebola virus-infected research monkeys were introduced into the facility. Richard Preston's account of the incident, *The Hot Zone,* inspired the film *Outbreak* in 1998. More recently, Ebola reemerged in the Democratic Republic of Congo in September 2007, resulting in at least 166 deaths.

Dengue fever sickened 1.2 million people in fifty-six countries in 1998. Reports of the bacterium *E. coli* O157:H7 frequent the news, as in 2007 when a cluster of people became ill. Health officials detected the organisms in hamburger patties, resulting in the recall of nearly twenty-two million pounds of ground beef to avert a nationwide outbreak. Can you imagine a mound of hamburger meat that size? Peanut butter contaminated with *Salmonella* caused over 425 cases of *Salmonella* infection spread over forty-eight states in February 2007, resulting in recalls of particular brands of the product. (Parents packing peanut butter sandwiches into their children's lunch boxes were dismayed and at a loss to find an appropriate substitute.) Lyme disease, severe acute respiratory syndrome (known as SARS), and avian influenza are three more examples from a long list of new, emerging, and reemerging infectious diseases (TABLE 7.1); no nation can afford to be complacent regarding its vulnerability.

The Institute of Medicine is a branch of the National Academy of Sciences that advises the government on policy matters pertaining to the health of the public. In a 1992 report, *Emerging Infections: Microbial Threats to Health in the United States,* the institute defined emerging infections as "new, reemerging or drug-resistant infections whose incidence in humans has increased within the past two decades or whose incidence threatens to increase in the near future."

Despite the tremendous progress in the latter half of the twentieth century in controlling infectious diseases, including the eradication of smallpox, the introduction of antibiotics in the 1940s, improvement in sanitation, and an increase in the diseases for which immunization is available, the battle against infectious diseases is uphill (BOX 7.1). David Satcher, former director of the U.S. Centers for

---

TABLE 7.1 Examples of New, Emerging, and Reemerging Infections

Bacterial diseases	Protozoan diseases
Lyme disease	Cryptosporidiosis
Ehrlichiosis	Malaria
Escherichia coli O157:H7	Babesiosis
Legionnaires' disease	**Fungal diseases**
Tuberculosis	Coccidioidomycosis
Viral diseases	*Pneumocystis* pneumonia
Hantavirus pulmonary syndrome	
Ebola hemorrhagic fever	
Dengue fever	
Rabies	
Sever acute respiratory syndrome (SARS)	
West Nile	

BOX 7.1 Quotations Relating to Health and Infectious Disease

The world may have only a decade or two to make optimal use of the many medicines presently available to stop infectious diseases. We are literally in a race against time to bring levels of infectious disease down worldwide, before the disease wears the drugs down first.

—David Heymann, Executive Director, World Health Organization's Communicable Disease Program, 2000

Germs come by stealth
And ruin health
So listen, pard,
Just drop a card
To a man who'll clean up your yard
And that will hit the old germs hard.

—Sinclair Lewis, *Arrowsmith*

Everyone has the right to a standard of living adequate for the health and wellbeing of himself and of his family including food, clothing, housing, and medical care. . . .

—Article 25, Universal Declaration of Human Rights, adopted by the General Assembly of the United Nations, December 10, 1948

Health is a state of complete physical, mental, and social wellbeing and not merely the absence of disease or infirmity. . . .

—Constitution of WHO, July 22, 1946

Infectious disease is one of the few genuine adventures left in the world. The dragons are all dead and the lance grows rusty in the chimney corner. . . . About the only sporting proposition that remains unimpaired by the relentless domestication of a once free-living human species is the war against those ferocious little fellow creatures, which lurk in the dark corners and stalk us in the bodies of rats, mice and all kinds of domestic animals; which fly and crawl with the insects, and waylay us in our food and drink and even in our love.

—Hans Zinsser, *Rats, Lice, and History,* 1935

It is time to strengthen our research efforts . . . so that we can unlock the mysteries behind antibiotic resistance and discover new scientific weapons in the battle to detect and control emerging infectious diseases.

—Albert Gore, former vice president of the United States, June 1996

Ingenuity, knowledge, and organization alter but cannot cancel humanity's vulnerability to invasion by parasitic forms of life. Infectious diseases which antedated the emergence of humankind will last as long as humanity itself and will surely remain, as it has been hitherto, one of the fundamental parameters and determinants of human history.

—William H. McNeill, *Plagues and Peoples,* 1976

Pathogenic microbes can be resilient, dangerous foes. Although it is impossible to predict their individual emergence in time and place, we can be confident that new microbial diseases will emerge.

—Institute of Medicine, *Emerging Infections: Microbial Threats to Health in the United States,* 1992

The microbe that felled one child in a distant continent yesterday can reach yours today and seed a global pandemic tomorrow. Pitted against microbial genes, we have mainly our wits.

—Joshua Lederberg, 1958 Nobel Prize winner, 1988

On a good day, we hold them at bay. On a bad day, they're winning. Our task is a lot like trying to swim against the current of a raging river.

—Michael Osterholm, former Minnesota state epidemiologist and founder of an infectious disease control company, 2000

Disease Control and Prevention (CDC) and later surgeon general of the United States, warned that "our ability to detect, contain, and prevent emerging infectious diseases is in jeopardy."

The International Red Cross warned of the danger of infectious diseases in a report published on June 6, 2000. The report spoke of "the silent tragedy" of deteriorating health services and the death of thirteen million people from preventable diseases, primarily infectious in nature, in the previous year. Further, compared with floods and earthquakes, which grab news headlines and donors' cash, the uncontrolled spread of disease steals far more lives. Over 150 million people have died of AIDS, TB, and malaria alone since 1945, compared with the more than twenty-three million lives lost in wars. In a recent year, 160 times more people died from AIDS, malaria, respiratory diseases, and diarrhea than were killed in that year's natural disasters, including the massive earthquakes in Turkey, floods in Venezuela, and cyclones in India. Malaria kills one million people a year, mostly children. According to the World Health Organization (WHO), a child dies of malaria every thirty seconds. TB has been on the upswing in North Korea; statistics from that country reveal that five million of its twenty-two million people are infected.

The WHO reported in 2008 that infectious diseases were the second leading cause of death worldwide, resulting in one-third of the fifty-nine million deaths occurring worldwide each year (FIGURE 7.2). Almost ten million children under the age of five die each year, and their leading killers are infectious diseases: pneumonia, diarrhea, malaria, measles, and HIV. These figures are an underestimate, because surveillance and reporting networks are woefully deficient in many less-developed countries. The leading infectious killers in the world, according to the WHO, include bacterial, viral, protozoan, and worm diseases. Initially, one would attribute these devastating statistics to the poverty associated with developing nations, but, as surprising as it may seem, in the United States infectious diseases remain in the top ten causes of death (FIGURE 7.3). No wonder the director-general of WHO stated in a 1996 report, "We stand on the brink of a global crisis in infectious diseases. No country is safe from them. No country can any longer afford to ignore this threat."

So what's the bottom line? What grade would the world now be awarded in terms of its success in coping with microbial diseases? Certainly, under the leadership of the United Nations, WHO, the CDC, and other organizations, the burden of infectious diseases around the world can be lessened and a higher grade achieved.

In fact, during the 1950s, 1960s, and 1970s microbial diseases appeared to be on their way out. It was a span of years heralded by optimism and progress in public health. The first polio vaccine was introduced by Jonas Salk in the 1950s and ushered in a time of successful mass vaccination campaigns. Fewer than one thousand polio cases occurred in 1967 in western Europe and North America as compared with over seventy-five thousand in 1955. It appeared that malaria would be taken off the list of diseases "of major importance," according to WHO and the Pan-American Sanitary Conference. The times were good; as the economy

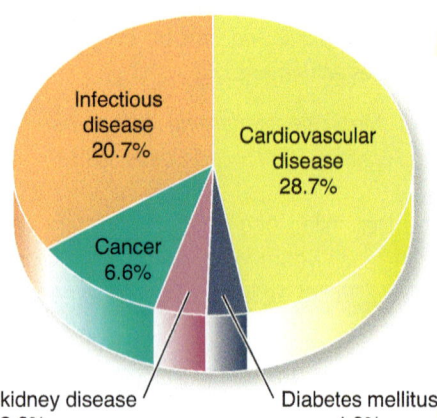

Infectious disease 20.7%

Cardiovascular disease 28.7%

Cancer 6.6%

Liver and kidney disease 2.6%

Diabetes mellitus 1.9%

FIGURE 7.2 The five leading causes of death from disease. There were 59.4 million deaths worldwide in 2007. Cancers and cardiovascular, respiratory, and digestive diseases can also be caused by infections, and thus the percentage of deaths due to infectious diseases may be even higher than shown. *Source:* World Health Organization, *World Health Statistics,* 2008.

of nations improved, poverty decreased, and so did the burden of microbial diseases. A 1966 CDC report, in what might be considered an address on the state of the union's health, glowed with the promise of the conquest of microbial diseases. William H. Stewart, surgeon general of the United States in 1967, told a gathering of health officers that it was "time to close the book on infectious diseases and shift all national efforts to chronic diseases." Stewart's optimism was echoed by health officials in other developed nations of the world.

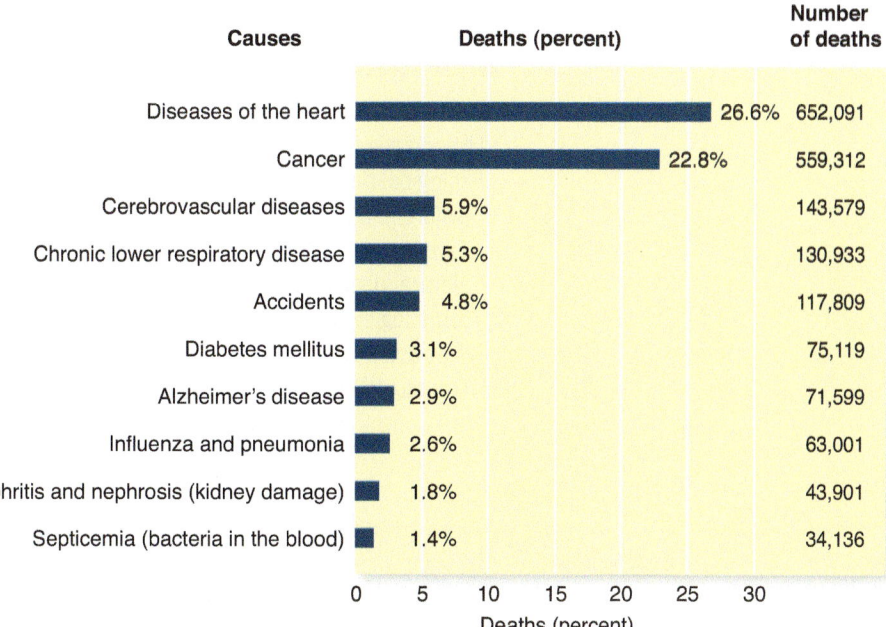

FIGURE 7.3 Leading causes of death in the United States. Adapted from *Deaths: Final Data for 2005, National Vital Statistics Reports,* Vol. 56, No. 10, April 24, 2008, National Center for Health Statistics/CDC.

Unfortunately, the warnings were not heeded and the optimism was not shared in developing countries—countries that constitute a large part of the world's population. From 1980 to 1992 alone, the CDC reported a 22% increase in infectious diseases (excluding AIDS). Data presented in a 1996 article published by the *Journal of the American Medical Association* indicated a greater than 50% increase in deaths caused by microbes in the United States since 1980. Despite the tremendous strides in infectious disease control over the past century, data from the U.S. National Center for Health Statistics indicate that microbial disease remains as a leading cause of death in the United States (Figure 7.3) But the scientific community was slow to acknowledge that the bubble of antisepsis and disease control was about to burst.

Why were new diseases emerging and older ones reemerging, with a vengeance, as it sometimes appeared? The 1992 Institute of Medicine's report, *Emerging Infections: Microbial Threats to Health in the United States,* warned that microbes were winning the battle and that our previous complacency and optimism had weakened our ability to counterattack. Essentially, it appeared that the choreography of adaptation between microbes and humans was beginning to come apart at the seams because of a variety of linked and overlapping factors considered below (TABLE 7.2): world population growth, urbanization, ecological disturbances, technological advances, microbial evolution and adaptation, and human behavior.

Factors Responsible for Emerging Infections

World Population Growth

By the end of 2012 more than seven billion people will be living on the earth, according to U.S. Census Bureau estimates. The growth rate is estimated at 1.25%, which means that by 2050 the population will have soared to over nine billion

TABLE 7.2 Factors Responsible for Emerging Diseases	
World population growth	Microbial evolution and adaptation
Urbanization	Antimicrobial resistance
Ecological disturbances	Evasive strategies
Deforestation	Human behavior and attitudes
Climatic changes	Complacency
Natural disasters (drought, floods)	Migration
Technological advances	Societal factors
Air travel	
Transfusion of unsafe blood	

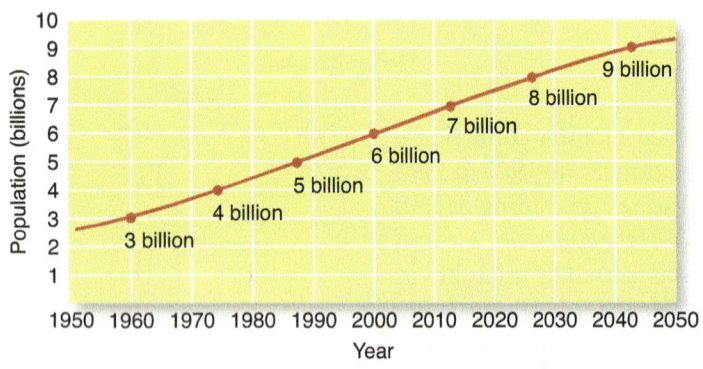

FIGURE 7.4 World population, 1950–2050. Projections are based on an estimated annual growth rate of 1.25%. *Source:* U.S. Census Bureau, *International Data Base,* July 2007 version.

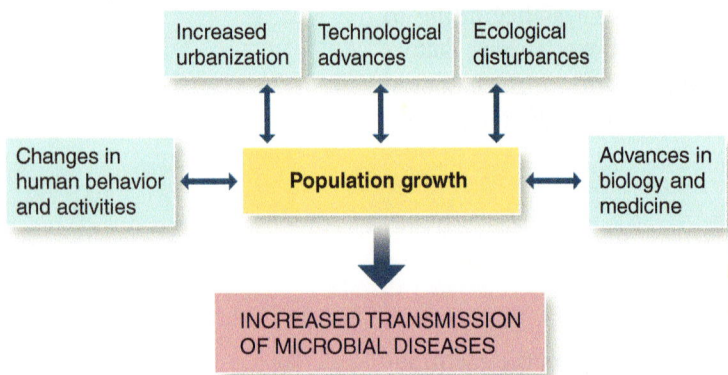

FIGURE 7.5 Population explosion: the "hub" of the problem.

(FIGURE 7.4). To add to the problem of the burgeoning population, 80% of the population is living in less-developed countries (of which 60% are tropical and subtropical areas) with a diminished capacity to cope with population increase. Several factors have been cited for the current crisis of new and emerging infectious diseases, but the population explosion is central to the issue (FIGURE 7.5).

Thomas Malthus' (1776–1834) *An Essay on the Principle of Population as It Affects the Future Improvement of Society* warned of the negative influence that unchecked population growth could have on societies, primarily because of inadequate supplies of food. (Charles Darwin's insights into evolution and the process of natural selection were strongly influenced by Malthus.) It appears that Malthus was right, as evidenced by famine in Africa and in other parts of the world over the past 100 or more years. But there is another negative consequence of overpopulation: transmission of infectious diseases. The total population of a country or a region, in itself, is not as crucial as its population density—the number of people per square mile in a defined area. Person-to-person transmission is facilitated as population density increases. Other diseases are carried by **biological vectors**, including mosquitoes, ticks, lice, and flies, and in some cases **zoonotic diseases** are transmitted from animal to human. Whatever the mode of transmission, population density is a significant factor. Consider, for example, a classroom with fixed dimensions and assume that one person in the class has a cold, but there are only ten other students randomly spaced throughout the room; on the other hand, consider the same classroom with sixty students. Clearly, the chain of transmission is fostered in the larger population, simply because respiratory droplets are able to traverse the shorter distance from contact to contact when the population density is higher.

The age distribution of the population is also of considerable significance in terms of the risk factors. For example, in the United States people over age sixty-five constituted 14.4% of the population in 2007—a figure that is expected to reach almost 28% by 2050. Elderly populations are more susceptible to microbial diseases, presumably because the strength of their immune system has declined. They serve as an increasing source of infection for family and community members. The point is that predictors of infectious disease need to take into account not only the total population and population density but also the demographics of age distribution in that population.

As the world population increases, there are a number of consequences that foster an increase in infectious diseases. For example, Dhaka, the capital city of Bangladesh, has many slum areas. Bangladesh is the world's seventh most populous country, with approximately 150 million people. Population control is Bangladesh's most urgent problem, along with the attendant low per capita income. As would be expected, high levels of malnutrition exist, with much of the population getting less than one-third of the normal food intake because agricultural production has not been able to keep pace with population growth. The consequences in terms of infectious diseases, particularly diarrhea, due to poverty and poverty-related conditions are dramatic.

Urbanization

Zaynab Begum lives in Bangladesh in a Dhaka slum, along with her husband and three children in a primitive hut, less than 6 square meters in size, constructed of bamboo and makeshift materials. There is no electricity, running water, or toilet, and an open sewer runs outside the hut. Zaynab's husband is a rickshaw puller; her fate is shared by millions of others who have left their villages and migrated to the cities in search of work and a better life.

The trend to urbanization dates back to early in the twentieth century. In 1900 just 13% of the world's population lived in urban areas. According to the 2007 UN *World Urbanization Prospects* report, the percentage has increased to 50%, and is expected to climb to 60% by 2030 (FIGURE 7.6). The world's ten largest urban areas are listed in TABLE 7.3; about half of these megacities are in developing countries. The magnitude of the effect of urbanization on communicable diseases varies dramatically in developed and developing countries, as a function of the economy and the public health infrastructure necessary to cope with the stress of increasing population density. The challenge of maintaining acceptable standards of sanitation and hygiene is far more difficult in developing countries, but it is also important to keep in mind that pockets of poverty and despair also exist in the United States and other developed nations.

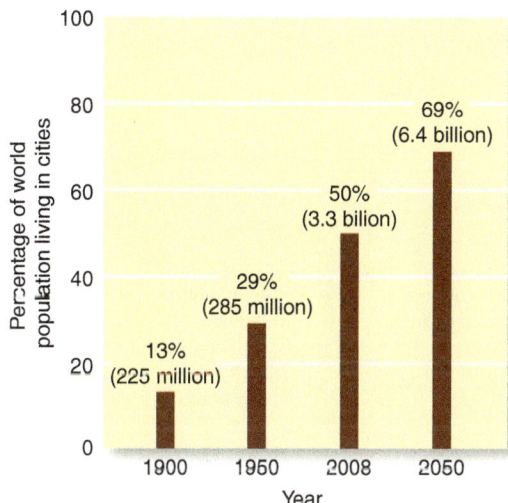

FIGURE 7.6 Progressive urbanization of our planet. Adapted from United Nations, Department of Economic and Social Affairs/Population Division, *World Urbanization Prospects: The 2007 Revision*, 2008.

TABLE 7.3 World's Ten Largest Urban Agglomerations	
Tokyo, Japan	Karachi, Pakistan
Mumbai, India	Mexico City, Mexico
Lagos, Nigeria	New York City, USA
Dhaka, Bangladesh	Jakarta, Indonesia
São Paulo, Brazil	Calcutta, India

The rankings vary depending on definition of urban agglomerations and yearly estimates of current population.

Urbanization frequently leads to poverty and together set up a cycle of infectious disease (FIGURE 7.7). In July 2000 at a meeting of the G-8 countries (the top seven industrialized countries, plus Russia) in Okinawa, the nations pledged to break the vicious circle of poverty suffered by citizens of developing countries. Sub-Saharan Africa is home to 68% of the world's 33.2 million people living with HIV (FIGURE 7.8) because of a variety of factors, an important one of which is poverty-associated urbanization. The drain on natural resources, including safe drinking water, is excessive, whereas at the same time problems of pollution, including human waste disposal and sanitation, are magnified. Untreated human waste by the tons are dumped into the rivers, streams, and oceans. Ultimately, slums and shantytowns develop. The United Nations estimates half of the population on the African continent—482 million people—live in slums. The people live in filth and squalor (FIGURE 7.9); rodent populations increase as sanitation decreases, and the cycle of disease is perpetuated. Rodents may harbor fleas, which transmit a variety of diseases, including the Black Death of fourteenth-century Europe, now known simply as the Plague.

AUTHOR'S NOTE
I have been to Japan on a few occasions, the first being during my service as a young military officer in the U.S. Army stationed just outside of Japan. I am always impressed with its cleanliness, despite the ever-present crowds.

In April 2000 the CDC concluded in a five-year study that city dwellers get sick more often than their rural counterparts and that people living in poverty are sick more often. Upton Sinclair's 1906 novel, *The Jungle*, portrayed the unsanitary practices and working conditions, especially for the workers, in the Chicago meatpacking industry (BOX 7.2). Prevention of communicable diseases is a major component of public health and is more problematic in cities than in rural areas and wide open spaces.

On the other hand, it does not necessarily follow that disease runs rampant in the megacities. Consider, for example, the Tokyo-Yokohama area, ranked as the world's largest urban area (Table 7.3), which is hardly poverty stricken or disease ridden. In fact, population statistics indicate that the Japanese people enjoy the longest life expectancy. The country's economy and public health infrastructure make it possible for them to cope with urbanization. By contrast, in most developing countries the crush of humanity

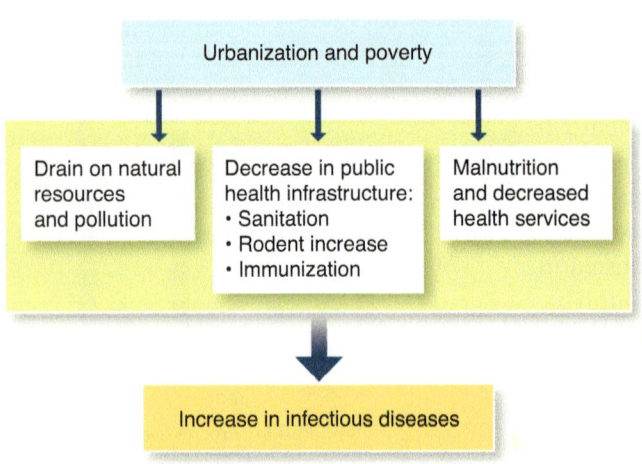

FIGURE 7.7 Relationships among poverty, urbanization, and infectious disease.

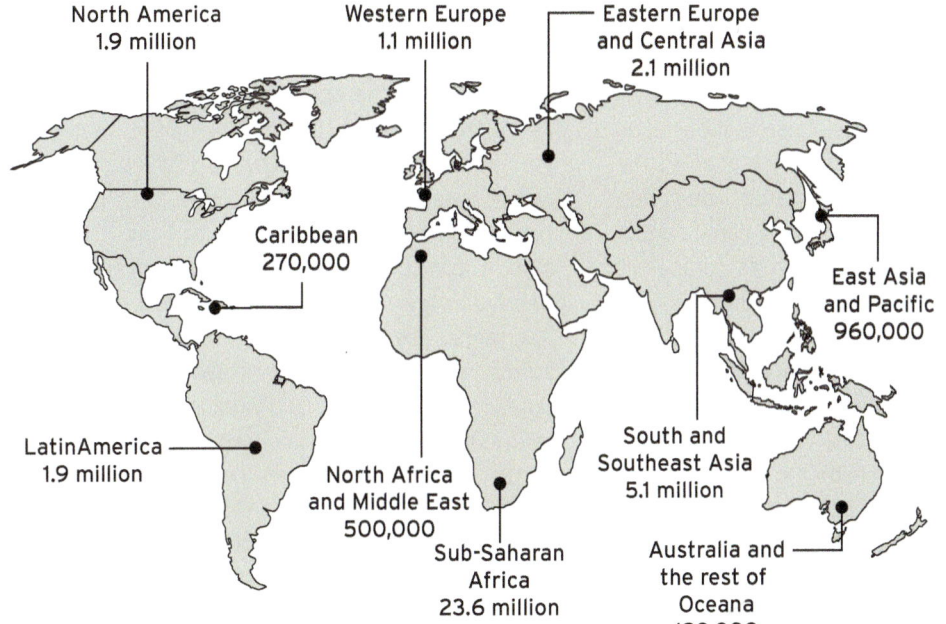

North America
1.9 million

Western Europe
1.1 million

Eastern Europe
and Central Asia
2.1 million

Caribbean
270,000

East Asia
and Pacific
960,000

LatinAmerica
1.9 million

North Africa
and Middle East
500,000

South and
Southeast Asia
5.1 million

Sub-Saharan
Africa
23.6 million

Australia and
the rest of
Oceana
120,000

FIGURE 7.8 Global distribution of HIV. Adapted from UN program on HIV/AIDS, 2005.

(a)

(b)

FIGURE 7.9. Slums and shantytowns. Poverty is associated with a lack of sanitary facilities, an increase in rodent populations, a lack of safe drinking water, and other circumstances that contribute to infectious diseases. **(a)** A shack in rural Panama. Author's photo. **(b)** A slum area in Manila, Philippines. © RubberBall/Alamy Images.

BOX 7.2   An excerpt from *The Jungle,* Upton Sinclair

There was another interesting set of statistics that a person might have gathered in Packingtown—those of the various afflictions of the workers. . . . There were the men in the pickle-rooms, for instance, where old Antanas had gotten his death; scarce a one of these that had not some spot of horror on his person. Let a man so much as scrape his finger pushing a truck in the pickle-rooms, and he might have a sore that would put him out of the world; all the joints in his fingers might be eaten by the acid, one by one. Of the butchers and floorsmen, the beef-boners and trimmers, and all those who used knives, you could scarcely find a person who had the use of his thumb; time and time again the base of it had been slashed, till it was a mere lump of flesh against which the man pressed the knife to hold it. The hands of these men would be criss-crossed with cuts, until you could no longer pretend to count them or to trace them. They would have no nails,—they had worn them off pulling hides; their knuckles were swollen so that their fingers spread out like a fan. There were men who worked in the cooking-rooms, in the midst of steam and sickening odors, by artificial light; in these rooms the germs of tuberculosis might live for two years, but the supply was renewed every hour. There were the beef-luggers, who carried two-hundred-pound quarters into the refrigerator-cars; a fearful kind of work, that began at four o'clock in the morning, and that wore out the most powerful men in a few years. There were those who worked in the chilling-rooms, and whose special disease was rheumatism; the time-limit that a man could work in the chilling-rooms was said to be five years. There were the woolpluckers, whose hands went to pieces even sooner than the hands of the pickle-men; for the pelts of the sheep had to be painted with acid to loosen the wool, and then the pluckers had to pull out this wool with their bare hands, till the acid had eaten their fingers off. There were those who made the tins for the canned-meat; and their hands, too, were a maze of cuts, and each cut represented a chance for blood-poisoning.

and the tide of urbanization are overwhelming and beyond the financial resources necessary to construct sewage systems and to develop and maintain a public health infrastructure. Laura Garrett, in *The Coming Plague,* refers to cities as "microbe magnets" and "microbe heavens." "Graveyards of mankind" is a term used by British biologist John Cairns.

Ecological Disturbances

Deforestation

Almost half of the earth's forests either no longer exist or have been damaged, possibly to the point of no return, as a result of agriculture, settlement, logging, and mining over the past eight thousand years. Deforestation is a major factor in the eruption of emerging and reemerging diseases. Wilderness habitats serve as reservoirs for a large variety of insects and other animals that harbor infectious agents. When the village or town becomes too crowded, whether it is in a poor and developing country or in a developed country, expansion occurs into the surrounding areas for tracts of land on which to build. Generally, the first event to give notice that construction is about to take place is the whine of the chain saws signaling deforestation followed by bulldozers moving in to uproot the tree stumps (FIGURE 7.10). Every time a tree is felled or a bulldozer digs up the soil to create another shopping or housing development, microbes and other organisms are displaced. Fungi and bacteria and their spores are released into the environment and may alight and colonize on a human or animal and possibly give rise to a new or reemerging disease. Examples of outbreaks of certain fungal diseases have been reported in construction workers, particularly in the southwest. Perhaps this is what Louis Pasteur was referring to 150 years ago when he advised, "the microbe is nothing; the terrain is everything."

Human intrusion into the environment fosters contact with wildlife and with insects and plays a major role in the migration of these displaced species into villages, communities, and backyards in search of food. An example is the rise of rabies in the eastern part of the United States as a result of rabies-infected raccoons foraging for food in the garbage cans of suburban and rural communities. Chagas disease is a protozoan disease carried by beetles, commonly called kissing bugs, because they bite on the face and lips where the skin is thin. They are particularly prevalent in Brazil and other areas of South America. In the early 1900s construction of the Central Railroad in Brazil was undertaken through the heavily forested tropical wilderness, a project that necessitated large-scale deforestation. You can guess the outcome—the indigenous mammals were displaced, as were the beetles that fed on them for their blood meal. Humans and their domesticated animals took up the slack and became infected, as did rodents; the latter conveyed the disease to species of beetles that inhabit housing in urban populations.

In 1998 and 1999 a new and deadly virus, named "Nipah," killed more than one hundred people in Malaysia after first showing up on a pig farm. It is speculated that the pigs ate dropped fruits infected with the virus, which was then spread to farm workers. Fruit bats, also known as "flying foxes," are the world's largest bats and have been identified as the natural reservoir. In the years preceding the outbreak, massive deforestation took place and scores of fruit trees were destroyed causing the bats to forage elsewhere, including that remote pig farm surrounded by fruit trees.

Leishmaniasis, a protozoan disease carried by infected sand flies, is a striking example of the consequence of deforestation and the emergence of urban disease. The disease, once limited to mammals of the forest, is now urban, primarily as a result of deforestation. The circumstances are similar to those described for Chagas disease (FIGURE 7.11)

Although it can be an attempt to relieve suffering and death, the intrusion of humans into ecosystems can backfire. Inadequate assessment of the public health impact can inadvertently increase microbial disease. This was the case in the construction of Egypt's one billion dollar Aswan High Dam, a ten-year project completed in 1970. The dam harnessed the uncontrolled Nile River by creating Lake Nasser. (Gamal Abdel Nasser was the Egyptian president from 1956 to 1970.)

FIGURE 7.10   Deforestation. As people move into areas that were formerly forests, there is increased contact with animals, including insects that harbor infectious microbes. Further, the displaced animals return to neighborhoods that were once their lands in search of food. Author's photo.

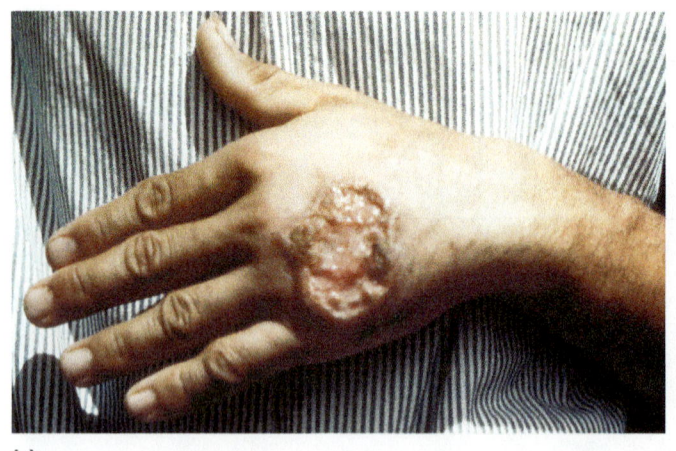
(a)

(b)

FIGURE 7.11 Leishmaniasis is a proto-zoan infection transmitted by infected sand flies. **(a)** A leishmaniasis skin ulcer on the hand of a Central American villager. Courtesy of Dr. D. S. Martin/CDC. **(b)** Primitive living conditions in a village in Central America. It occupies an area that was formerly a forest and is encircled by a perimeter of trees. Sand flies are poor fliers but can traverse the short distance from their forest habitat. Author's photo.

AUTHOR'S NOTE

In the spring of 1999 I spent six weeks in Salvador, Brazil at the Institute for Tropical Medicine in completion of the requirements of the M.P.H. degree at the Harvard School of Public Health to study tropical diseases in the natural context of their host. The last week of my stay was in the Amazon and included a visit to a small village with a high incidence of leishmaniasis. It was readily apparent why this was the case; the village bordered a heavy forest (Figure 7.11). Trees had been felled to allow for the construction of primitive dwellings and yet, less than a mile away, leishmaniasis was not prevalent. The reason—sand flies are poor fliers and could not fly far from the forest. As a result of the deforestation, the village habitants and their dogs provided a blood meal for the sand flies.

Unfortunately, the walls of the dam served as a new and convenient habitat for snails. The population of snails boomed, resulting in an increase in the incidence of schistosomiasis, a disease caused by a parasitic worm with a complicated life cycle requiring snails for its completion.

Rift valley fever, a viral hemorrhagic disease carried by mosquitoes, is another example of the downside of the Aswan High Dam project. An epidemic of this viral disease occurred close to the dam area, resulting in illness in 200,000 people and over 500 deaths. The epidemic was the result of a thriving mosquito population in the flood lands created by the dam.

The most contemporary example of the potential consequences of humans' intrusion into the forest is that of AIDS. Most scientists agree that the origin of human AIDS is the result of the simian immunodeficiency virus (SIV) that made the species leap from infected chimpanzees and sooty mangabeys to humans (FIGURE 7.12).

Climactic Changes

What about the effect of climate on the emergence of microbial diseases? There is ample evidence that global warming and climatic changes cause ecological disturbances that affect the incidence and distribution of infectious diseases (TABLE 7.4). The twentieth century witnessed an increase in average global temperature attributed largely to the burning of fuels and forests, resulting in an increase in

FIGURE 7.12 The interspecies leap. AIDS, which originated in Africa, is presumed to have jumped the species barrier from infected chimpanzee or sooty mangabeys to humans. Other infectious diseases of humans have made a leap from animals to humans. © Jan van der Hoeven/ShutterStock, Inc.

TABLE 7.4 Infectious Diseases Linked to Climatic Changes

Disease	Biological Agent	Transmission
Malaria[a]	Protozoan	Mosquitoes
Rift valley fever[a]	Virus	Mosquitoes
Hantavirus[a]	Virus	Mice
Cholera[a]	Bacterium	Waterborne
E. coli infection	Bacterium	Waterborne
Cryptosporidiosis	Protozoan	Waterborne
Hepatitis	Virus	Waterborne
Leptospirosis	Bacterium	Waterborne
Lyme disease	Bacterium	Ticks
Dengue (breakbone) fever	Virus	Mosquitoes

[a]Directly related to El Niño.

carbon dioxide and other heat-trapping greenhouse gases. The effects are seen not only in human health but also in the disruption of ecosystems and the resulting interference with food productivity. A meeting of world leaders was held in Kyoto, Japan in 1997 to develop countermeasures against the impending threat; these talks resulted in the Kyoto Protocol.

The following year, 1998, was the warmest year worldwide since 1880 when fairly accurate recordings began. The first two months were dominated by a record-breaking El Niño-influenced weather pattern, with wetter than normal conditions across much of the southern third of the United States and warmer than normal conditions across much of the northern two-thirds of the country. The increasingly high temperature exacerbated the extreme regional weather and climate anomalies associated with El Niño. That year brought into focus what scientists had long hypothesized, namely, that global warming could favor outbreaks of a variety of infectious diseases. Furthermore, seven of the eight warmest years on record have occurred since 2001.

In the case of vector-borne diseases, the vector or the microbe, or both, may be influenced by the temperature. TABLE 7.5 summarizes data on tropical diseases and indicates the likelihood of alteration in their distribution as a result of climate change. Malaria, a protozoan disease transmitted by mosquitoes, is at the top of the list; an increase in both temperature and rainfall extends habitats favorable to mosquitoes. (On the other hand, increased temperature and decreased rainfall favor the distribution of sand flies, the vectors responsible for transmission of leishmaniasis, a protozoan disease.) Estimates are that an increase in mean ambient temperature in central Africa by 2°C would extend the range of the vectors of sleeping sickness, filariasis, and leishmaniasis, allowing for these diseases of the tropics to invade marginal temperature zones. Further, higher temperatures may push malaria transmission to higher altitudes, causing epidemics, as has occurred in

TABLE 7.5 Status of Major Vector-Borne Diseases and Predicted Sensitivity to Climate Change

Disease	Population at Risk, in Millions[a]	Prevalence of Infection	Present Distribution	Possible Change of Distribution as a Result of Climatic Change
Malaria	2,400	300-350 million	Tropics, subtropics	Highly likely
Dengue	1,800	10-30 million	All tropical countries	Very likely
Schistosomiasis	600	200 million	Tropics, subtropics	Very likely
Onchocerciasis	123	17.5 million	Africa, Latin America	Very likely
Lymphatic filariases	1,100	117 million	Tropics, subtropics	Likely
Yellow fever	450	More than 5,000 cases	Tropics, S. America, Africa	Likely
Leishmaniasis	350	12 million infected, 500,000 new cases per year	Asia, southern Europe, Africa, Americas	Likely
African trypanosomiasis	55	250,000–300,000	Tropical Africa	Likely

[a]Based on a world population estimate of six billion.
Adapted from Vital Climate Change Graphics. UN Environment Programme (UNEP) and GRID-Arendal 2000.

the highlands of Ethiopia and Madagascar. In Rwanda in late 1987, malaria incidence increased by 337% over the previous three-year period as a result of increases in temperature and rainfall. In the last decade the reported cases of malaria (the form of malaria with the highest fatality rate) in the North-West Frontier Province of Pakistan rose from a few hundred in 1983 to more than twenty-five thousand in 1990. This dramatic rise is attributed to unusually high temperatures at the end of the normal malaria season that extended the season. Malaria, tickborne encephalitis, and leishmaniasis (carried by sand flies) are on the upswing in Italy as a result of climate change according to an Italian environmental organization.

In 1993 hantavirus, the cause of a potentially fatal disease, emerged in the Four Corners area of the United States (where New Mexico, Utah, Colorado, and Arizona meet). The disease is transmitted by deer mice, the principal animal hosts of the virus, which feed on pine kernels. Higher than normal humidity favored an abundant crop of the pine kernels, which, in turn, led to a tenfold increase in the deer mouse population between 1992 and 1993. This is an excellent example of a climatic condition triggering a chain of events resulting in the emergence of an infectious agent.

Diseases in which infectious agents cycle through invertebrates to complete their development are particularly sensitive to subtle climate variations compared with diseases spread from human to human. Hence, it is imperative that consideration be given to and appropriate measures be enacted regarding the influence of global warming and other climatic changes on microbes and their vectors.

Human Disease and Prevention

Natural Disasters

Floods, hurricanes, earthquakes, drought, and the very rare tsunamis are environmental disturbances that place populations at risk of an increased burden of infectious diseases. In March and April of 2000 severe floods put the people of Mozambique and other southern African countries at risk for several diseases, particularly malaria and cholera. Up to 250,000 people in Mozambique alone were endangered by these two diseases. According to a WHO press release, "The threat of a malaria epidemic in the country is increasing and will be at its most dangerous in around three to six weeks time as flood waters gradually subside, the rain stops, and warm temperatures return—ideal breeding conditions for mosquitoes. . . . Before the floods, there were between six and ten cholera cases a week; since the floods it has increased to 120 cases per week." Myanmar (formerly Burma) was hit by tropical cyclone Nargis on May 2, 2008 resulting in over 100 thousand deaths because of heavy rains and 12-foot water surges unleashed by the storm. Mud slides were triggered, contaminating wells that were a source of drinking water and blocking latrines, raising pubic health concerns as a result of a breakdown in sanitation.

Drought is related to increased famine and results in an increase in infectious diseases, as has been witnessed in eastern Africa since early 1999, placing sixteen million people at risk. In Ethiopia alone, about eight million people are affected. The country is one of the five poorest countries in the world, with an average life expectancy of forty-three years, the fifth lowest in the world.

On December 6, 2004 the world was shocked to witness mountains of water cascading on the northwest coast of the island of Sumatra, Indonesia triggered by an earthquake measuring 9.2 (out of 10) on the Richter scale. Natives and tourists ran to high ground for their lives, but an estimated 230,000 didn't make it and died. Half a million people were displaced from their homes, and thousands remain unaccounted for. The rapid response and level of relief measures from the international community was unprecedented. Almost immediately, groups worked to prevent infectious disease epidemics by providing "clean" water and bed-nets, initiating a measles vaccination program, and working to prevent and treat soil-transmitted worm infections. Although gaps in the public health infrastructure of the area and in the management of catastrophic events were uncovered, no large-scale outbreaks of infectious disease occurred, and mortality from disease was lower than anticipated.

Another natural disaster occurred when Hurricanes Katrina and Rita inundated large sections of New Orleans and surrounding parishes within a month of each other, August 29 and September 24, 2005, respectively (FIGURE 7.13). The pictures on television and in the newspapers and magazines were horrific; thousands of desperate people crowded into the New Orleans Convention Center, whereas others clung atop trees and roof tops hoping for rescue from the swirling waters. Surprisingly, there were no major outbreaks of infectious disease, although there were cases of wound and gastrointestinal infection primarily due to exposure to contaminated flood waters. The major microbial culprit was mold. As the waters receded, mold thrived and grew in the high humidity and excess moisture. Anyone exposed risked respiratory infections. Further, the CDC reported the occurrence of

(a)

(b)

FIGURE 7.13 **(a)** A flooded neighborhood in New Orleans as a result of Hurricane Katrina. Courtesy of Jocelyn Augustino/FEMA. **(b)** A New Orleans resident inspects mold damage. Courtesy of Andrea Booher/FEMA.

eighteen cases of wound-associated illness caused by two species of *Vibrio*, five of which resulted in deaths (**FIGURE 7.14**). These infections generally result when open wounds are exposed to warm seawater containing specific vibrios; those with weakened immune systems and the elderly are particularly at risk.

Life in less-developed countries is a struggle against poverty and disease. Approximately 1.5 billion people do not have access to safe drinking water. The lack of food and water takes its toll on the maintenance of a healthy immune system and leads to high child mortality rates. Natural catastrophic events exacerbate the potential for microbial-caused diseases.

FIGURE 7.14 **(a)** An open wound on a hand. © Jonathan Noden-Wilkinson/ShutterStock, Inc. **(b)** A diagnostic culture of *Vibrio cholerae*, the cause of cholera. Courtesy of CDC.

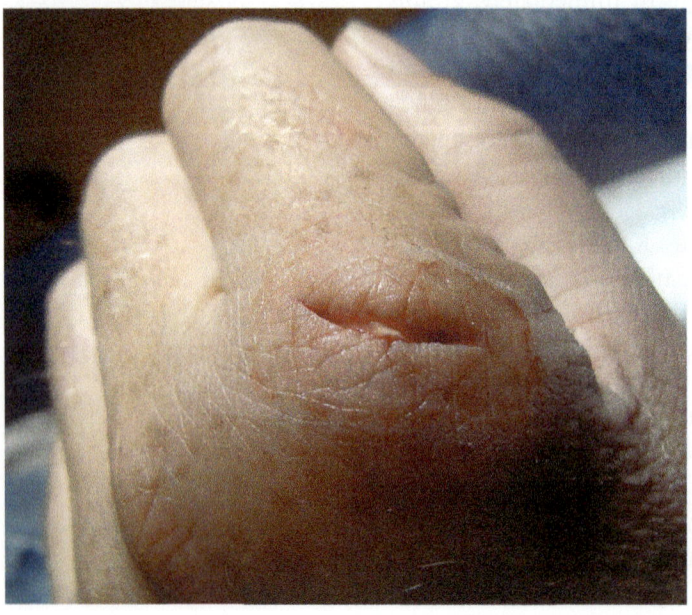

(a)

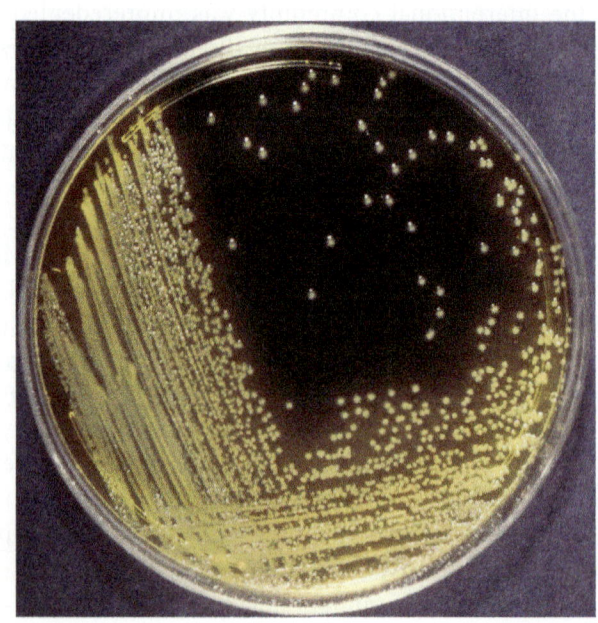

(b)

TABLE 7.6 Jet Travel: Microbes Without Passports

Approximate Flying Time From New York City

Sydney, Australia: 22 hours (1 stop)

Tokyo, Japan: 14 hours (nonstop)

Tel Aviv, Israel: 10 hours (nonstop)

Nairobi, Kenya: 16 hours (1 stop)

Incubation Period for Selected Diseases

Whooping cough: 7–10 days

Gonorrhea: 2–6 days

Salmonella food poisoning: 8–48 hours

Ebola fever: 4–16 days

Measles: 12–32 days

Chicken pox: 10–23 days

Technological Advances

Human activities lead to technological advances that may pose public health risks; jet travel is an example. It has been well documented that air travel plays a significant role in the transmittance of infectious diseases from continent to continent. TABLE 7.6 lists approximate flying times from New York City to distant places. The farthest destination is Sydney, Australia, taking twenty-two hours, less than the incubation time for many microbial diseases. This means that an infectious traveler could board a jet and arrive at any world destination in less than the time it takes for that passenger to show symptoms (FIGURE 7.15). Such an incident occurred in the spring of 2000: A tourist left Tel Aviv bound for Newark International Airport, an approximately ten-hour nonstop flight, and died of bacterial meningitis approximately two hours after landing. Fortunately, there were no reports of other passengers acquiring the disease. Bacterial meningitis has an incubation period of only a few hours to about two days.

FIGURE 7.15 A flight departure board at an international airport. Jet aircraft, a major technological advance of the twentieth century, serve as vectors for microbes around the world. © Neale Cousland/ShutterStock, Inc.

Not so lucky were thirteen travelers infected by a passenger with TB on a flight from Russia to New York. More recently, an international TB scare occurred in May 2007 when an Atlanta man previously diagnosed with an extremely drug-resistant strain of TB (XDRTB) traveled abroad. Although he did have TB, the diagnosis of XDRTB proved to be false. The good news is that none of his fellow passengers on the aircraft became infected. In these cases the best that can be done is to notify other passengers to seek medical advice.

Now consider the implications of the Airbus A380—a recently introduced and the world's largest commercial airliner—a super jumbo jet, double-decker, four engine craft with a wing span almost as big as a football field. In a three-class seating configuration it can carry 555 passengers, but in a one-class economy seating configuration approximately 850 can be accommodated. From an epidemiological point of view the aircraft is a nightmare—it serves as a huge potential mechanical vector capable of bringing infected people to any part of the world.

The infected people can carry the microbes to many different, final destinations, and in this way an epidemic can be triggered.

Microbes can be harbored and transported across borders not only in their human hosts but also in their baggage and personal items. Further, vectors harboring infectious agents can also travel; fleas can be carried in rugs transported by jet cargo from the Middle East and Asia. Public health officials inspect many items being transported from country to country and are authorized to impose quarantine in an effort to minimize the risks.

The use of whole blood and blood products is a life-giving and lifesaving practice. Unfortunately, in some countries blood and blood products may be hazardous to your health. According to WHO, most of the countries with an unsafe blood supply are developing nations, in which the chances of acquiring infectious diseases are highest. As population pressure increases, so does the demand for blood. In some countries blood is not screened and may harbor the causative agents of HIV infection, hepatitis, syphilis, malaria, and trypanosomiasis. Only recently has the blood supply in the United States been screened for the trypanosome protozoan parasite that causes Chagas disease. Although the disease is primarily spread by the bite of an infected beetle, also referred to as a "kissing bug," it can also be spread by blood transfusions and organ transplants.

To some extent advances in medical technology that make organ transplantation possible contribute to the burden of infectious diseases. Recipients are at increased risk for infection because they are on a regimen of immunosuppressive drugs to minimize organ rejection. Other conditions leading to immunosuppression include AIDS, certain inherited diseases, and malnutrition.

Prostate cancer is the second leading cause of cancer-related deaths in the United States. Transrectal ultrasound-guided biopsies of the prostate gland are common diagnostic procedures. According to the CDC, 624,000 procedures are performed annually. On July 21, 2006 the CDC reported on four cases of infection caused by *Pseudomonas aeruginosa* after transrectal ultrasound procedures. The infections were caused by contamination of the biopsy equipment that had not been properly sterilized. The bacterial strains recovered from patients matched the strains recovered from the lumen of the biopsy needle. This is an excellent example of a **nosocomial infection** (hospital-acquired infection).

Microbial Evolution and Adaptation

The 1940s ushered in the dawn of antibiotics—agents that were rightfully called "wonder drugs." Penicillin was the first, and numerous others quickly followed; some were tailored to be effective against a broad spectrum of bacteria, whereas others were more specific. It should be emphasized that antibiotics are not effective against viruses and hence should not be prescribed for viral infections. The number of lives saved worldwide over the past fifty years because of antibiotic therapy is beyond estimation. An individual today who is infected with a variety of life-threatening bacteria has a fighting chance, assuming antibiotics are administered promptly, whereas an individual infected fifty or sixty years ago had little chance of recovery. The development of antibiotics was a major factor leading to the optimism of the 1970s. Many dread diseases, so it seemed, were about to become vanquished. But it turns out that the tables are turning—antibiotics are losing their punch, and increasing numbers of microbes are resistant. The expression "I'm resistant to such and such an antibiotic" has no meaning; people do not become resistant to antibiotics—their microbes do.

Emblazoned on the cover of the September 12, 1994 issue of *Time* is the headline "Revenge of the Killer Microbes" and the question "Are we losing the war against infectious diseases?" (TABLE 7.7). Resistance to antimicrobial agents is at a crisis level worldwide (FIGURE 7.16). Vancomycin, an antibiotic considered by many to be the last stronghold in certain situations, is no longer effective against many bacterial strains that responded ten years ago. Some refer to antibiotic-resistant bacteria as "super bugs." What's happening? To put it in a nutshell, the forces of natural selection are in play. Antibiotics have been grossly misused and have promoted the emergence of antibiotic-resistant organisms in a Darwinian fashion. The antibiotic-resistant strains are the result of chance mutations, and their survival is favored by the presence of antibiotics. Antibiotics are the "selecting" and not the "causing" agent.

The battle against the natural process of microbial adaptation and change, whether exhibited by resistance against antibiotics or by evasive strategies, is an ever-present and ongoing struggle for survival. Failure to meet the challenge affords microbes the upper hand.

Insect vectors are also able to adapt to a changing environment. Malaria, a mosquito-borne protozoan disease, was thought to be a disease of the past, thanks to the application of the insecticide dichlorodiphenyltrichloroethane (DDT). Little did scientists realize that the forces of natural selection would again interfere as a result of the misuse of the insecticide. DDT-resistant mosquitoes emerged with a vengeance, and other vectorborne diseases shared their triumph over DDT. West Nile virus, a mosquito-borne agent, now threatens all of the contiguous United States, prompting ground and aerial spraying. Can insecticide-resistant mosquitoes carrying West Nile virus emerge?

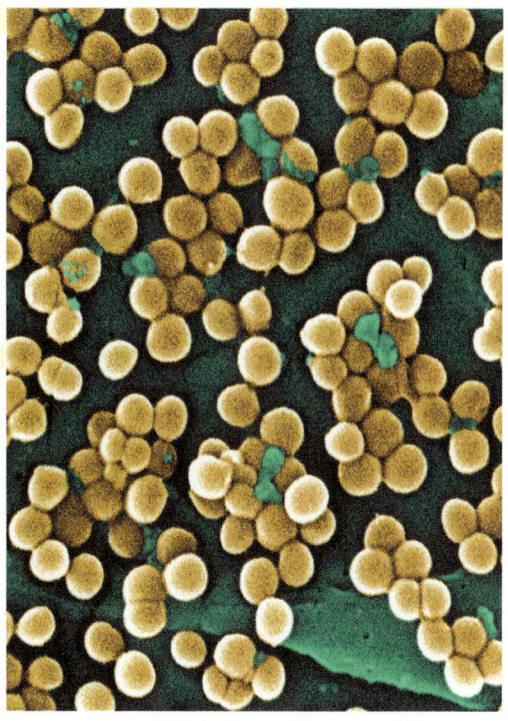

FIGURE 7.16 A scanning electron micrograph of methicillin-resistant *Staphylococcus aureus*, commonly referred to as MRSA. Courtesy of Janice Haney Carr/Jeff Hageman, M.H.S./CDC.

TABLE 7.7 Examples of Drug-Resistant Diseases		
Bacterial Disease	**Viral Disease**	**Protozoan Disease**
Tuberculosis	HIV infection	Malaria
Gonorrhea	Hepatitis B	Visceral leishmaniasis
Staphylococcal infection		
Shigellosis		
Typhoid fever		
Pneumococcal pneumonia		
Enterococcal infection		

Human Behavior and Attitudes

Complacency

How easy it is to cut corners on health-related matters when it appears that progress and improvement have taken place, leading to the false assumption that prevention and control are no longer necessary. Complacency is the belief that "it can't happen to me." The failure of people to complete their full dose of antibiotics because they are feeling better is a prime example.

A dramatic example of complacency is evident in the threatened resurgence of AIDS, particularly among young gay men, because of a return to risky sexual behavior fueled by glowing reports of new drug therapies for the management of AIDS. At least five million Americans have sex and/or drug habits that put them at high risk for acquiring AIDS. The number of cases in the United States has fallen dramatically since the peak of the 1980s; the decrease is primarily attributed to safer sex habits and avoidance of dirty needles by drug abusers, but public health officials worry that the decrease in cases could cause complacency and result in an increase in the number of cases as people return to unsafe sex practices.

People have become complacent about receiving immunization shots or keeping their immunization boosters up to date. According to the CDC only 80% of two-year-olds in the United States have been given the full sequence of currently recommended immunizations, primarily because of parents' complacency and concern about safety. As an example, consider that in 2005 only about 81% of children between the ages of nineteen to thirty-five months were fully immunized in the state of New York. This could mean trouble down the line. Past history reveals that a 10% decline in measles vaccination between 1989 and 1991 resulted in an outbreak of 55,000 cases, several thousand hospitalizations, and 120 deaths, indicating the power of immunization. Fortunately, children entering school are required to have proof of being up to date on their immunizations; students entering college must also have proof of being fully immunized. Nevertheless, some slip through the cracks. Individuals traveling to foreign countries need to be aware that particular immunizations may be necessary against diseases prevalent in that area. For example, in 1996 tourists traveling to yellow-fever areas neglected to be immunized against the disease and were responsible for infecting others with the disease upon their return to the United States and Switzerland.

Human Migration

Human migration is a major factor in the emergence and reemergence of many communicable diseases. The Population Reference Bureau estimates that in the mid-2000s about 191 million people lived outside their native countries. Populations on the move contribute to the emergence of disease beyond that resulting from voluntary urbanization fueled by a search for a better life. Population movement is frequently not a matter of choice but rather a forced movement because of wars and conflicts resulting from political upheavals. The United Nations defines **internally displaced persons** (**IDPs**) as "Persons or groups of persons who have been forced or obliged to flee or to leave their homes or places of habitual residence, in particular as a result of or in order to avoid the effects of armed conflict, situations of generalized violence, violations of human rights or natural or human made disasters, and who have not crossed an internationally recognized State border." The term *refugee* is reserved for those who are forced under the same circumstances to cross an international border. The Office of the UN High Commissioner for Refugees estimates the number of forcibly displaced persons at nearly thirty-three million worldwide. These people carry with them their microbes and microbial vectors, resulting in an exchange with intermingling populations. Malaria is an excellent example; refugees migrating through regions where malaria is endemic can acquire the infection and disseminate the disease to other areas. Malaria is a common cause of death among refugees in numerous countries, including Thailand, Somalia, Rwanda, the Democratic Republic of the Congo, and Tanzania.

Masses of people are forced to settle in uninhabitable environments without adequate shelter, food, clean drinking water, and latrines. Personal hygiene and sanitation may be virtually nonexistent, and what few facilities are available become quickly overwhelmed. People live in filth and squalor. These camps are hotbeds for epidemics, and their potential spreads as refugees continue to flee from one area to another (FIGURE 7.17). The ongoing Darfur conflict in western Sudan is a human catastrophe and a worst-case scenario. The United Nations has estimated 200 thousand to 400 thousand have died from violence and disease and another 2.5 million have been displaced to refugee camps.

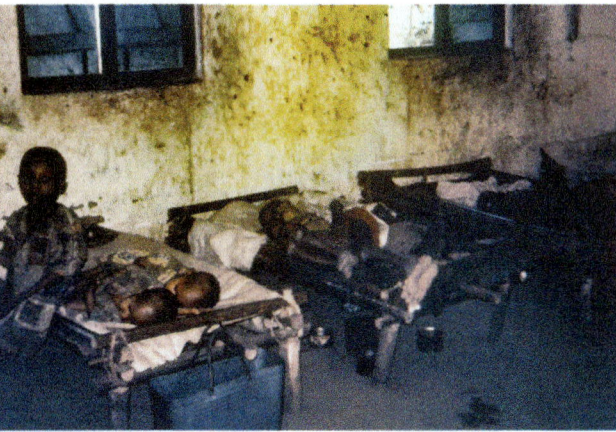

FIGURE 7.17 A refugee camp. Refugee camps are hotbeds of infection. Crowding and lack of hygiene and sanitation favor the incidence and transmission of disease. Courtesy of Dr. Lyle Conrad/CDC.

Wars and civil unrest, in addition to creating refugees and displaced persons, disrupt the public health infrastructure and favor the spread of disease. The destruction of housing leads to increased human-to-human and human-to-vector contact; decline in water management programs and a lack of treatment facilities are contributory factors.

Societal Factors

In many societies, particularly in developed countries, family life and structure have changed as a result of economic growth and increased opportunities for women. In most American families both parents work, leading to an increase in child care centers. Millions of children attend day care centers, which put them at risk for a variety of intestinal parasites, diarrhea, middle ear infections, and meningitis. For example, outbreaks of shigellosis, a diarrheal disease, have caused problems in many day care centers around the country. (To a large extent, the simple act of hand washing by the staff after they change a diaper is an effective control measure.) Children convey the microbes to their family members, many of whom in turn bring their microbes to the workplace.

As longevity increases, so does the number of elderly citizens requiring nursing homes, day care centers, and assisted living environments. Like child day care centers, these facilities are potential hotbeds for the emergence and spread of communicable diseases within the resident population and the staff, visitors, and their contacts.

Food production and dietary habits also affect the spread of microbial diseases. Globalization of the food supply, centralized processing, fast-food

(a)

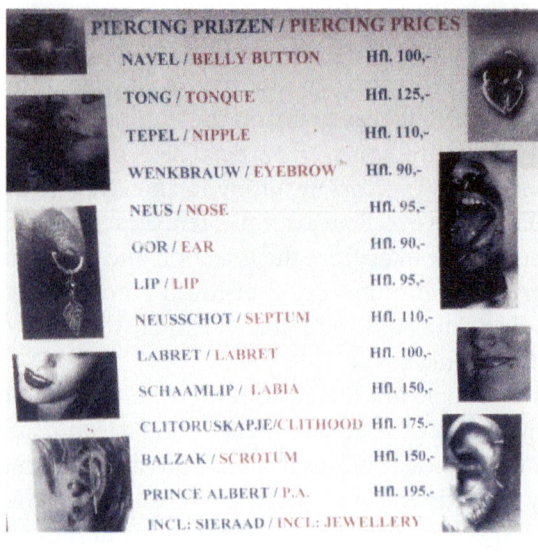

PIERCING PRIJZEN / PIERCING PRICES	
NAVEL / BELLY BUTTON	Hfl. 100,-
TONG / TONQUE	Hfl. 125,-
TEPEL / NIPPLE	Hfl. 110,-
WENKBRAUW / EYEBROW	Hfl. 90,-
NEUS / NOSE	Hfl. 95,-
OOR / EAR	Hfl. 90,-
LIP / LIP	Hfl. 95,-
NEUSSCHOT / SEPTUM	Hfl. 110,-
LABRET / LABRET	Hfl. 100,-
SCHAAMLIP / LABIA	Hfl. 150,-
CLITORUSKAPJE / CLITHOOD	Hfl. 175,-
BALZAK / SCROTUM	Hfl. 150,-
PRINCE ALBERT / P.A.	Hfl. 195,-
INCL: SIERAAD / INCL: JEWELLERY	

(b)

FIGURE 7.18 Tattooing and body piercing. Tattooing and body piercing are a risky part of popular culture. The skin is invaded, potentially resulting in serious infections because of the use of unclean instruments. (a) Tattoos. © Photodisc. (b) Anything goes if the price is right (Amsterdam). Author's photo.

restaurants, dining out, and take-out food are all significant. Foodborne diseases are a major public health problem in the United States. For example, an outbreak of hepatitis A in March and April 1997 resulted from children eating contaminated strawberries in school lunches and was traced to a food processing company in California. In 1994 and 1995 a salmonella-contaminated snack food manufactured in Israel caused illness in Israel, the United Kingdom, and the United States. In 1997 raspberries from Guatemala contaminated with a pathogenic protozoan parasite were responsible for cyclosporiasis, a food-associated illness, in the United States and Canada. During the spring and continuing into the summer of 2008, over one thousand cases of salmonellosis occurred in the United States, presumably due to certain varieties of contaminated tomatoes. Other recent examples of foodborne outbreaks were described earlier in the chapter. In a better economy and a family structure in which both parents work, many people rely more on prepared foods to reduce household chores. Fast-food restaurants and take-out restaurants are part of our social structure. Food has increasingly become a source of recreation. Consider, for example, a typical conversation: "So, what'll we do tonight?" Answer: "Let's eat out!"

This is all well and good, assuming that personal hygiene and sanitary control measures practiced by food handlers are not compromised. Television news shows have aired segments featuring high-end restaurants that are enough to make you sick!

Tattooing and body piercing are ancient art forms that have continued through the centuries. In developed countries these practices have long been popular with sailors and bikers. Young people in the 1990s and the new millennium have brought the trend into the mainstream. Tattoo and body piercing parlors are found in many countries, including the United States; for a price you can get just about any part of your body tattooed or pierced (FIGURE 7.18). The risk of infection with a variety of microbes, particularly staphylococci, is a real possibility, and patrons are often at risk because of nonsterile instruments and poorly trained personnel. Recently, the CDC reported forty-four cases of methicillin-resistant

Staphylococcus aureus skin infections in Ohio, Kentucky, and Vermont in 2004-2005 as a result of thirteen unlicensed tattooists, presumably because of the use of nonsterile equipment in these three states. Even if the establishment is certified by a local health authority, let the buyer beware!

■ Overview

This chapter makes the case that despite the optimism of forty or fifty years ago, microbial diseases have not been eliminated (with the single exception of small-pox) but rather flourish as a major cause of mortality and morbidity around the world. The reasons for this are based on world population growth, urbanization, ecological disturbances, technological advances, microbial evolution and adaptation, and human behavior. A quotation from Donald A. Henderson during his tenure as associate director of the U.S. Office of Science and Technology Policy serves as an excellent way to close this chapter:

> The recent emergence of AIDS and dengue hemorrhagic infections, among others, [is] serving usefully to disturb our ill-founded complacency about infectious diseases. Such complacency has prevailed in this country [USA] throughout much of my career. . . . It is evident now, as it should have been then that mutation and change are facts of nature, that the world is increasingly interdependent, and that human health and survival will be challenged, ad infinitum, by new mutant microbes, with unpredictable pathophysiological manifestations.

■ Self-Evaluation

PART I: Choose the single best answer.

1. World population is now estimated at
   a. 8 billion **b.** 6.6 billion **c.** 4.6 billion **d.** more than 8 billion
2. Which one of the following people warned of unchecked population growth?
   a. Satcher **b.** Stewart **c.** Malthus **d.** Darwin
3. In the United States, by 2050 estimates are that people over the age of sixty-five will constitute about what percentage of the population?
   a. 20% **b.** 28% **c.** 32% **d.** not predictable
4. The construction of the Central Railroad in Brazil led to an increase in
   a. leishmaniasis **b.** malaria **c.** tuberculosis **d.** Chagas disease
5. Which disease has been shown most conclusively to be linked to climate change?
   a. leishmaniasis **b.** *E. coli* infection **c.** leptospirosis **d.** malaria
6. Tattooing carries a risk of infection. Which organism (or disease) is most likely to be involved?
   a. staphylococci **b.** *E. coli* **c.** meningitis **d.** cryptosporidiosis

PART II: Fill in the blank.

1. In the United States infectious diseases are the leading cause of death. True or false? _____

2. Name an infectious disease considered to be in the "top ten" worldwide. _____

3. According to the U.S. Census Bureau, the world's largest urban area is _____.

4. Diseases transferred from animals to humans are called _____.

5. What does the acronym IDP stand for? _____

6. Diseases transmitted from animals to humans are called _____ diseases.

PART III: Answer the following.

1. List five reasons why infections are emerging and increasing.

2. Choose two of the reasons you gave in question no. 1 and discuss them.

3. Why was the Aswan Dam a "disaster"?

4. Cairns described cities as "graveyards of mankind." Explain.

5. A number of quotations are cited in this chapter. Develop your own quotation that targets the problem of new, emerging, and reemerging infections.

6. Do you believe that a grade of C- should be given to the world for its efforts in coping with microbial diseases. What grade would you award, and why?

7. In 1967 the surgeon general of the United States declared it was "time to close the book on infectious diseases," but events proved otherwise. What was the basis of the surgeon general's remark?

8. Complacency is listed as a major factor responsible for the continued threat of microbial diseases. Describe some specific examples, including examples for which you and family members may be "guilty."

Control of Microbial Diseases

It is a disturbing fact that Western Civilization which claims to have achieved the highest standards of health in history, finds itself compelled to spend ever-increasing sums for the control of disease.

— Rene Dubos, 1987

■ Preview

Advances in public health during the twentieth century have decreased the burden of microbial disease on a worldwide basis, particularly in the United States and other industrialized nations. Sanitation and clean water, food safety, immunization, and antibiotics are major factors in the control of microbial diseases. Each of these factors is discussed in this chapter.

Is the general health of your generation better than that of your parents' or grandparents' generations? Most decidedly, your response would be "yes," despite the current problem of new and reemerging infections. Society, particularly in developed countries, is no longer "plagued by plagues," but thousands of citizens in the United States and in other developed areas live in pockets of poverty and disease similar to the developing world. Even in developing countries the burden of disease has been reduced, although certainly not to the degree enjoyed by the richer nations of the world.

TABLE 8.1	Ten Great Public Health Achievements in the United States, 1900–1999[a]

Vaccination[b]

Motor vehicle safety

Safer workplaces

Control of infectious diseases (includes sanitation hygiene, antibiotics, and clean water)[b]

Decline in deaths from coronary heart disease and stroke

Safer and healthier foods[b]

Healthier mothers and babies

Family planning

Fluoridation of drinking water

Recognition of tobacco as a health hazard

[a]Not ranked in order of importance.
[b]Focused on reduction of infectious diseases.

Adapted from CDC. *Morbidity and Mortality Weekly Report* 48 (1999):241–243.

Consider that a person born in the United States in the early 1900s could anticipate an average life span of forty-five years and that the death rate at birth was slightly higher than 10%. Back then tuberculosis was the leading cause of death. Today, life expectancy in the United States, as estimated by the Central Intelligence Agency *Worldfact Book,* is 78.2 years (75.15 for males and 80.97 for females), an increase of thirty-three years. The overall worldwide life expectancy is 66.26 years according to the 2008 *Worldfact Book.* Life expectancy ranges from a high of 84.33 years in Macau to a low of 32.23 in Swaziland. Surprisingly, the United States ranks number forty-five in a list of 221 countries despite the fact that it spends the most on health. Wealthy Americans are among the world's healthiest people, whereas Americans on the bottom rungs have a life expectancy characteristic of sub-Saharan Africa.

An eminent historian stated, "The retreat of the great lethal diseases was due more to urban improvements, superior nutrition, and public health than to curative medicine." In 1999 the Centers for Disease Control and Prevention (CDC) published a report entitled *Ten Great Public Health Achievements in the United States, 1900–1999* (TABLE 8.1); three of those achievements focused directly on a reduction of infectious diseases (FIGURE 8.1). This chapter considers these achievements under the following sections: "Sanitation and Clean Water," "Food Safety," "Immunization," and "Antibiotics."

Sanitation and Clean Water

Part of the daily routine each morning as you prepare for the day is attending to matters of sanitation and personal hygiene, including showering, brushing your teeth, and using a clean flush toilet. These activities are taken for granted, but they are luxuries. On the other side of the globe in a poverty-stricken and war-torn

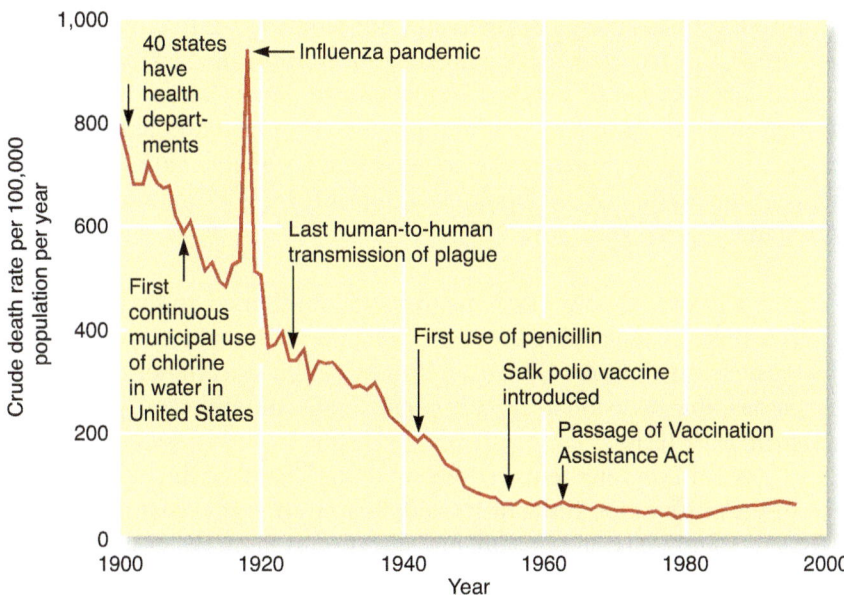

FIGURE 8.1 Major public health events and their influence on crude death rate, 1900–1996. Infectious diseases were a major cause of morbidity and mortality in the early 1900s. Public health intervention focused on improvements in sanitation and hygiene, implementation of antibiotics, and immunization programs. These strategies were effective over the generations. Adapted from CDC, *Morbidity and Mortality Weekly Report* 48 (1999): 621–629; adapted from G.L. Armstrong, et al., *JAMA* 281 (1999): 61–66. Date of first continuous municipal water chlorination is from American Water Works Association. *Water Chlorination Principles and Practices.* AWWA manual M20, Denver, CO: AWWA, 1973.

part of Indonesia, refugees occupy makeshift shelters in camps and live under miserable conditions with inadequate sanitation, food, and water. A newspaper account tells the plight of one family: "Mr. _____ watched cockroaches crawl over his wife and three children asleep on the dirt floor of a tent they share with three other families. He lamented the lack of food and medicine to treat the diarrhea, measles, eye infections, flu, and other ailments that spread quickly here. There is nothing but misery and sorrow in this place."

According to the World Health Organization's (WHO's) Water Supply and Sanitation Collaborative Council, 2.6 billion of the world's citizens lack basic hygiene and sanitation facilities and more than 1.5 billion people in the world do not have access to a daily supply of clean water. Every year, about 1.5 million people die of diarrheal diseases because they lack the most basic sanitation.

World leaders agree that hygienic means of sanitation and a safe supply of drinking water are basic human needs. Kofi Annan, a past UN Secretary General, said it all in his words: "We shall not finally defeat AIDS, tuberculosis, malaria, or any of the other infectious diseases that plague the developing world until we have also won the battle for safe drinking water, sanitation and basic health care."

Development of Sanitation

Urbanization is not a new phenomenon but one that dates back millennia to the times when hunter–gatherers first saw a benefit in pooling their meager resources by living together in villages. These villages have evolved into today's cities. In Chapter 7 population growth and the resulting urbanization in the mid-nineteenth century were cited as major factors contributing to the challenge of infectious disease control. The germ theory and the idea of contagion had yet to be developed, and urban centers were struggling to establish infrastructure to keep pace with the burgeoning masses. Little regard was paid to public health measures. Industrial development and immigration led to an influx into the cities, which, in turn, fueled poor housing, overcrowding, lack of clean water, and lack of facilities for disposal of human waste. John Cairns, a British biologist, termed cities the "graveyards of mankind." The nineteenth-century outbreaks of cholera in London illustrate the consequences of inadequate sanitation and hygiene; Arno Karlen, in his book, *Man and Microbes* (Simon & Schuster, 1996), wrote the following:

> The city's seven sewer systems were uncoordinated and relied on defective pipes. They received tons of human and animal feces, dead animals, waste from abattoirs [slaughter houses], effluvia from hospitals and tanneries, the occasional human corpse, and contaminated ground water from cemeteries.
>
> All of London's refuse ended up in the Thames, which provided most of the city's water. Cholera arrived; there were eight separate water companies and just one experimental filtration system. Water not taken from the Thames came from wells, many as badly polluted as the river. The city drank, cooked, and washed in its own filth. With the crowding and dirt, once a waterborne disease was established, further person to person transmission was virtually assured.

London was not the only city to be so afflicted. Filth and squalor prevailed in Europe and around the world. Cities in the United States were hardly models of cleanliness and sanitation. Sewage disposal systems were few, and outhouses, overflowing cesspools, and garbage-littered streets flourished, as did tuberculosis, diphtheria, scarlet fever, and typhoid fever. Somewhere about the middle of the nineteenth century a sanitary reform movement gradually arose from the ashes of human corpses, debris, and human and animal wastes, perhaps fired by the third cholera epidemic to hit London (on the heels of the second epidemic). The combination of disease, filth, and lack of shelter (FIGURE 8.2) led to the enactment of laws relating to sanitation, including sewage and water treatment, garbage collection, and other public health measures. In the 1880s Louis Pasteur and Robert Koch triumphed with their discoveries. The germ theory of disease was established, and microbes were at last linked to sanitation and (microbial) disease.

The germ theory was embraced in Europe and in the United States. Sanitation was "in," and sanitary engineers and bacteriologists (a term that preceded "microbiologists") flourished (FIGURE 8.3). In 1887 the Marine Hospital Service was established and charged with monitoring cholera in immigrants on ships coming into New York. This facility was the forerun-

FIGURE 8.2 Factors in the spread and emergence of infectious disease. **(a)** People living in squalor without adequate shelter and basic sanitation are at increased risk for infectious disease. © Vishal Shah/ShutterStock, Inc. **(b)** Despite major advances over the past century, substandard levels of living such as those shown in this old drawing persist in underdeveloped countries and in pockets of developed countries. © National Library of Medicine.

(a)

(b)

Human Disease and Prevention

ner of today's National Institutes of Health. Other public health laboratories and organizations were established in major cities around the world, and bacteriologists and sanitary engineers worked in concert. Public health statutes promoting sanitation and good hygiene were passed and implemented, and by 1900 forty states had health departments.

Over the twentieth century the focus of health departments changed from meeting the urgent and immediate needs of basic sanitation to delivering health services (TABLE 8.2). The 1920s through the 1950s witnessed great strides in public health strategies to control infectious diseases (FIGURE 8.4). Great attention was paid to the construction of water and sewage treatment facilities, chlorination, better housing, control of tuberculosis and venereal diseases (now called sexually transmitted diseases or infections), food production and distribution, animal and pest control measures, and garbage disposal (FIGURE 8.5). The public was bombarded with

FIGURE 8.3 At the beginning of the 20th century, the age of sanitation began to emerge. © National Library of Medicine.

TABLE 8.2 Changing Role of Health Departments

Health department services in 1900 (driven by urgent needs)
  Sewer construction
  Water supply inspection
  Sewage disposal
  Nuisance and pest control
  Privy inspection or removal
  Milk supply inspection
  Infectious disease control (tuberculosis, diphtheria and croup, scarlet fever, smallpox, and typhoid fever)

Health department services in 1999 (developing an organized approach)
  Monitoring community health status to identify potential hazards
  Investigating disease outbreaks and safety hazards in the community
  Mobilizing community partnerships to solve health problems
  Developing policies and plans that support individual and community health efforts
  Enforcing laws and regulations that protect health and ensure safety
  Linking populations with needed personal health services and ensuring the provision of health care when otherwise unavailable
  Ensuring a competent public health and personal health care workforce
  Evaluating effectiveness, accessibility, and quality of personal and population-based health services
  Researching new ideas and innovative solutions to health problems

Source: CDC.

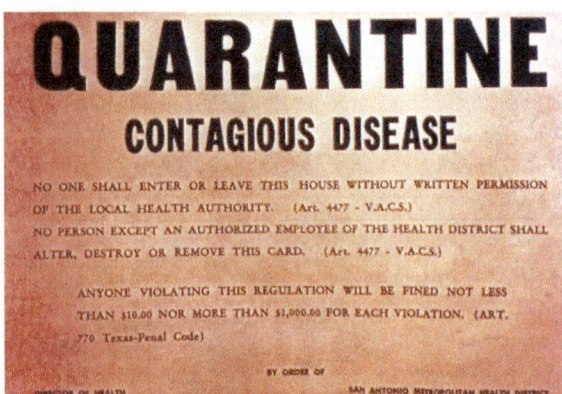

FIGURE 8.4 Public health statutes were passed and enforced by health officers. Quarantine signs were required by law to be placed on homes in which an individual was suffering from a "contagious" disease. Courtesy of San Antonio Metropolitan Health District. Used with permission.

FIGURE 8.5 Garbage disposal is an important public health measure. Garbage accumulation attracts rats and other rodents and animals that serve as reservoirs and vectors of infectious disease. Author's photo.

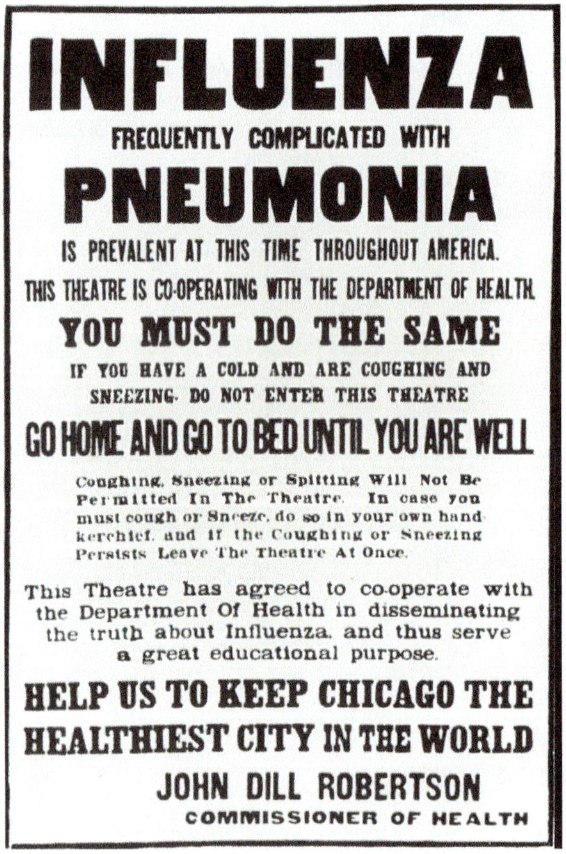

FIGURE 8.6 Departments of health fostered in the population an appreciation of good personal hygiene in an effort to control the dissemination of infectious disease. Reproduced from Robertson, J.D. A. *Report on an Epidemic of Influenza in the City of Chicago in the Fall of 1918*. Chicago, 1918.

information regarding the evil of "germs" and their transmission from the sick to the healthy by various modes of transmission (FIGURE 8.6). The "gospel of germs" was accepted, and rub-a-dub-dub, scrub, dust, and clean-clean-clean were heralded. Somewhere along the line the expression "cleanliness is next only to godliness" became a household dictate; some say this expression is attributed to Mahatma Gandhi of India as his "battle cry" during his efforts in the 1920s to 1930s to clean up the villages. These new efforts paid off. Malaria, plague, tuberculosis, and other diseases were markedly reduced; the last major outbreak of plague in the United States occurred during 1924 and 1925 in Los Angeles.

Human Waste Disposal

Your first impulse may be to laugh at learning there is actually a World Toilet Organization, founded in 2001 to promote sanitation. On the other hand, you'll recall that 2.6 billion people across the globe lack access to appropriate toilet facilities. Further, 200 million tons of human waste goes uncollected and untreated around the world because of the lack of toilets (FIGURE 8.7). World Toilet Day was celebrated on November 19, 2008.

The safe disposal of human excreta is central to sanitation, and its significance cannot be overemphasized. The General Assembly of the United Nations declared 2008 as an International Year of Sanitation in an effort to accelerate progress on worldwide improvements in sanitation in recognition of the fact that 2.6 billion people in the world lack proper sanitation facilities, particularly in developing countries and in

(a)

(b)

FIGURE 8.7 Human waste disposal. (a) Contamination of food and water by fecal material is a major cause of many infectious diseases. Author's photo. (b) Primitive and simple toilets are relatively inexpensive and efficient if properly constructed. Author's photo.

poverty pockets in developed countries. Estimates are that only 40% of the population have no choice but to squat and defecate in the open directly onto the ground.

In India, as an example, it is estimated that over 100 million households have no toilets and ten million households use buckets for waste disposal. Further, 900 million liters of urine and 135 million kilograms of fecal material need to be disposed of each day. About half a million children die every year in India because of dehydration resulting from diarrheal diseases that are frequently traceable to open defecation. The subject of toilets and defecation is hardly dinnertime conversation, but it is a fact of life and an integral part of the history of human hygiene.

Organizations such as the United Nations and the World Bank are working on the development of sewage disposal systems in developing countries. Bathrooms and flush toilets, such as our society is accustomed to, are not necessary goals; certainly, less luxurious and primitive facilities, be they outdoor or indoor, are affordable and effective (Figure 8.7). Programs to improve poor sanitation are in progress in slums and squatter settlements of the world's poorest countries. The Kampung Improvement Program in Indonesia, a highly successful program, has focused on covering open sanitation drains and on bringing reasonably clean water to families. The Orangi Pilot Project has reached 650,000 people in a poor neighborhood in Karachi, Pakistan. Orangi is the largest squatter settlement in Karachi, with approximately one million inhabitants. According to a public health official, "People don't need a flush toilet in every home or a faucet in every room. But with a standpipe for every three units, with adequate pit latrines and other forms of [waste] treatment, the services can be there and the health of the children maintained."

AUTHOR'S NOTE

In New Delhi, India there is the Museum of Toilets, which has a collection of artifacts, pictures, and objects illustrating the historical development of toilets since the year 2500 B.C. On display at the museum is a replica of the throne of King Louis XIII with its built-in commode he used while giving audience.

Clean Water

"Water, water everywhere, nor any drop to drink." This famous line from the classic 1798 poem, *The Rime of the Ancient Mariner,* by Samuel Taylor Coleridge, depicts the desperate plight of an old mariner surrounded by a sea of undrinkable water.

FIGURE 8.8 Clean water: a luxury. For much of the world's population, water is not piped into their homes and must be transported in containers. Bodies of water may serve for clothes washing, bathing, water for animals, and drinking water. © Marcus Brown/ShutterStock, Inc.

It can also serve to depict the desperation of one-fourth of the world's population that have only limited access to water that may or may not be safe (**FIGURE 8.8**). The WHO estimates that over one and half billion people worldwide lack access to clean water; in some villages people, primarily women and children, spend a major part of their day carrying buckets to a source of clean water. The UN Millennium Development Goals aims to reduce the number of people lacking access to water by 50% by the year 2015. Not surprisingly, there is a strong correlation between access to safe drinking water and child health (**FIGURE 8.9**).

In March 2007 the WHO's Water Supply and Sanitation Collaborative Council stated the following:

A lack of access to safe drinking water, inadequate sanitation facilities and poor hygiene strongly interferes with basic human development. Water-related diseases, including diarrhea, are a major cause of death amongst young children and each year they kill more children worldwide than HIV/AIDS. In India, child mortality due to HIV/AIDS counts for 0.7% while 20% die from diarrheal diseases. In Malawi, where 14% of the children under five years of age die due to HIV/AIDS, diarrheal diseases are a bigger killer and are responsible for 18% of children's deaths.

The WHO estimates that at any given time, perhaps one-half of all people in the developing world are suffering from one or more of the six main diseases

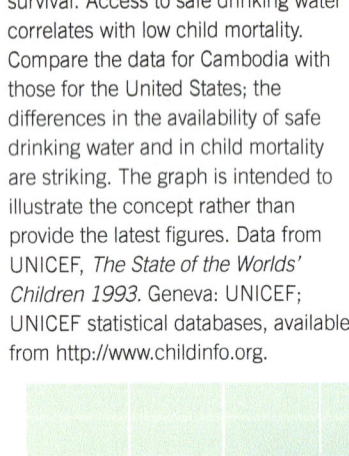

FIGURE 8.9 Water quality and child survival. Access to safe drinking water correlates with low child mortality. Compare the data for Cambodia with those for the United States; the differences in the availability of safe drinking water and in child mortality are striking. The graph is intended to illustrate the concept rather than provide the latest figures. Data from UNICEF, *The State of the Worlds' Children 1993*. Geneva: UNICEF; UNICEF statistical databases, available from http://www.childinfo.org.

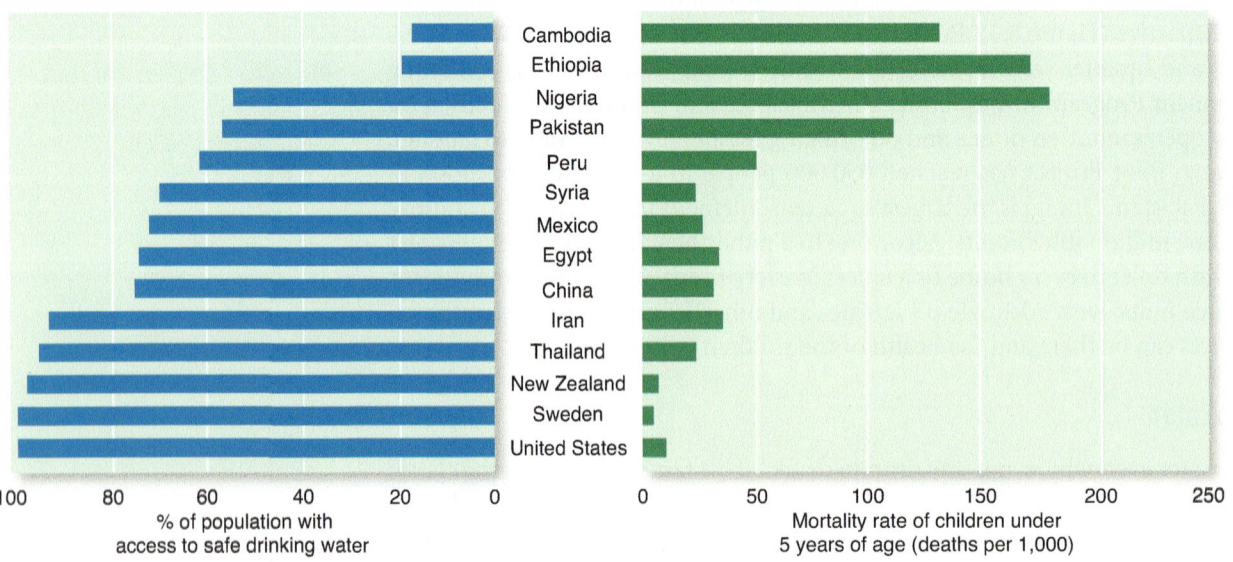

Human Disease and Prevention

associated with drinking contaminated water. The poorest people in the world are paying many times more than their richer compatriots for the water they need to live and are getting more than their share of deadly diseases because supplies are dangerously contaminated. Control of waterborne diseases is a particularly difficult problem when excreta are disposed of in a way that allows fecal material to gain access to water sources and food supplies.

Improvement of water quality was, and continues to be, a major public health priority aimed at the decline of waterborne and water-associated diseases. Anywhere from a 20% to 80% decline in morbidity and mortality is possible with improved water sanitation. In some countries people are infected with guinea worms, which bite their way through their victim's flesh, or with cholera so devastating that in several hours their life is threatened by severe dehydration resulting from massive loss of water through diarrhea. In the village of Chiladi in rural Brazil some end up with brown spots on their hands and serious symptoms due to drinking the arsenic-contaminated water that was supposed to be clean (BOX 8.1).

BOX 8.1   Arsenic in the Well and in the Wood

Sometimes, in an effort to alleviate a problem, well-meaning public health officials initiate interventions that backfire. Consider the following example.

Until about thirty years ago millions of people in Bangladesh and West Bengal, poor and densely populated regions, drank surface water contaminated with disease-producing microbes from shallow hand-dug wells, streams, and ponds resulting in a high burden of disease. To combat the high incidence of death and disease resulting from contaminated drinking surface water, international aid agencies such as UNICEF and local health officials installed tube wells to tap groundwater.

A tube well is a simple device constructed of steel pipes sunk deep into the ground fitted with a pump handle. You might find one at a roadside picnic area where piped water is not available. The pump is sealed topside to prevent water leaking back down the pipe. Microbes are filtered out as the groundwater trickles through the aquifer, resulting in microbiologically safe water.

An estimated 3.5 million wells gave millions of people in the area access to the groundwater. The water was expected to be the answer to the epidemics associated with the use of contaminated unsafe surface water. Although the number of waterborne diseases was markedly reduced, the price was too high; the groundwater was contaminated with naturally occurring arsenic in concentrations well above the accepted levels. By the mid-1990s thousands of people had been diagnosed with arsenic poisoning, and a new crisis existed.

Chemically, arsenic is categorized as a heavy metal, as is mercury and lead. It is usually excreted from the body, but if excess amounts are ingested it accumulates. Arsenic is very toxic and interferes with essential enzyme systems, resulting in death due to multiorgan failure. Historically, the use of arsenic as a poison for political assassinations dates back several centuries. Some historians believe that Napoleon was killed by food and beverage tainted with arsenic. ("Arsenic and Old Lace" was a hilarious and highly popular play that opened on Broadway in 1941. It is a comedy about two elderly sisters who poisoned lonely old men with elderberry tea containing arsenic, strychnine, and a "touch" of cyanide.) Interestingly, salvarsan, developed about 1909, one of the first drugs to treat syphilis, was an arsenic-containing compound.

It seems that the people of Bangladesh and West Bengal have unwittingly gone "from the frying pan into the fire." The choice may be between drinking arsenic-free, microbiologically contaminated water and drinking arsenic-contaminated, microbiologically safe water. Fortunately, new and alternative safe water options based on rendering surface water microbiologically safe and on treating arsenic-contaminated groundwater are increasingly available.

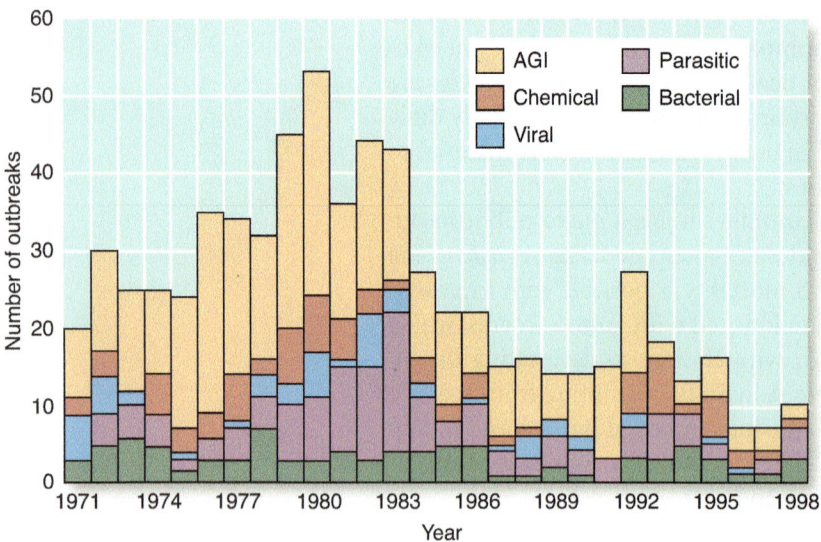

FIGURE 8.10 Number of waterborne disease outbreaks associated with drinking water in the United States, 1971–1998. Legislation to regulate the nation's drinking water over the past three decades has been successful. National, state, and local organizations and governments have worked together in this achievement. AGI, acute gastrointestinal illness of unknown etiology. Adapted from R. S. Barwick et al., *Morbidity and Mortality Weekly Report* 49 (2000):1–44.

However, even in developed countries vigilance needs to be maintained. A safe water supply can never be taken for granted, as the citizens of Milwaukee, Wisconsin, painfully discovered in 1993 when *Cryptosporidium parvum* caused the largest outbreak of waterborne disease in U.S. history. When developed countries experience fires, flood, hurricanes, wars, and other catastrophic events, they suffer the same misery of land and water becoming fouled with human and animal wastes and carcasses as the undeveloped countries. Hurricane Floyd hit North Carolina with a vengeance in September 1999 and the rivers filled with human waste, animal waste, and other pollution. Hurricane Katrina, one of the most monstrous storms over the past century, hit the U.S. Gulf Coast on August 29, 2005, and caused widespread damage over the area, particularly in New Orleans because of the resulting flooding. In the aftermath the population was exposed to contaminated drinking water and other public health hazards.

The CDC and the U.S. Environmental Protection Agency have maintained a collaborative surveillance system since 1971 and continue to report on the occurrence of waterborne diseases with the goal of characterizing and identifying the causative agents. Over the past century legislation has been implemented in an effort to regulate the nation's water supply. In 1972 the Clean Water Act came into effect as a response to the pollution of water in the United States with industrial and human wastes; bodies of water were becoming the nation's dumping ground. The goal was to reduce waterborne diseases and other adverse outcomes. Two years later, in December 1974, the Safe Drinking Water Act became effective and established measures to ensure the safety of drinking water at the tap. The Act was updated in 1986 and in 1996. As shown in FIGURE 8.10, the initiatives have paid off.

Food Safety

Safer foods are considered one of the ten great public health achievements of the United States during the twentieth century (Table 8.1). The WHO has expanded its food safety initiatives in response to new challenges, including health implications of genetically engineered foods. In the United States the Food Safety Inspection Service of the U.S. Department of Agriculture (USDA) is charged with the safety, labeling, and packaging of the nation's commercial supply of meat, poultry, and eggs. Nevertheless, outbreaks of foodborne disease continue to occur.

Leaders of the sanitation movement, initiated in the early 1900s, recognized that typhoid fever, tuberculosis, scarlet fever, botulism, and other diseases were transmitted by food (including milk) and water and advocated safer food-handling procedures, including pasteurization, refrigeration, hand washing, safer food processing, and pesticide application. *The Jungle*, a powerful 1906 novel by Upton Sinclair, portrayed the unsanitary practices in the Chicago meatpacking industry. Another excerpt from the book follows:

> There were some [cattle] . . . that had died, from what cause no one could say; and they were all to be disposed of here in darkness and silence. "Downers," the men called them; and the packing houses had a special elevator upon which they were raised to the killing beds, where the gang proceeded to handle them, with an air of business-like nonchalance which said plainer than any words that it was a matter of everyday routine. . . . and in the end Jurgis saw them go into the chilling rooms with the rest of the meat, being carefully scattered here and there so that they could not be identified.

The public reacted so strongly that within a year after publication of Sinclair's book the Pure Food and Drug Act was passed. Even today, news shows such as "Dateline" and "20/20" periodically air alarming stories about food processing. Despite the advances in the safety of the food supply, foodborne diseases remain a major cause of morbidity and mortality. Well over twenty-five microbes, including bacteria, viruses, protozoans, and worms, are causative agents of foodborne diseases. Further, changes in food-processing technologies, personal eating habits, and food distribution have contributed to the presence of new emergent pathogens. The CDC estimates the number of cases of foodborne diseases in the United States to be somewhere between 6.5 million and 76 million each year, with 325,000 hospitalizations and 5,000 to 9,000 deaths. More accurate estimates are not possible because of differences between states in disease reporting, data collection, diagnostic procedures, and analytical methods.

How does food become contaminated along the pathway from farm to plate? What are the sources of contamination? In 1999 the American Academy of Microbiology issued a report on food safety that cited seven practices in the food supply system that potentially contribute to the creation and persistence of situations that favor the likelihood of microbial contamination at levels high enough to cause disease; these seven practices are listed in TABLE 8.3.

Ultimately, foods are prepared and consumed in the home—the last stage in the journey from farm to plate. Cleanliness in the kitchen has received

TABLE 8.3 Practices Significant to Foodborne Diseases

Preslaughter practices	Agriculture
Preharvest practices	Food processing
Slaughtering practices	Food preparation and consumption in the home
Harvesting practices	

Modified from Stephanie Doores, Ph.D. *Food Safety: Current Status and Future Needs*. American Academy of Microbiology, 1999. Used with permission.

considerable attention in recent years. A biologist who calls himself the "Sultan of Slime" claims that "our bathrooms may be cleaner than our kitchens"; perhaps a course in "kitchen microbiology" is not a bad idea. Dish towels, sponges, counter surfaces, and hands are all culprits involved in the spread of foodborne pathogens.

New measures to increase food safety, some of which are "gimmicky," are in vogue. The (noninspiring) debate regarding the use of wooden versus plastic cutting boards still exists. One entrepreneurial company now advertises "cut and toss" disposable cutting boards. Toxin Alert, a Canadian company, has developed an "intelligent" food wrap, covered with antibody sensors, that changes color when it is contaminated with *Salmonella, Campylobacter, Escherichia coli,* and other bacteria. Pasteurized eggs are available and marketed under the brand name Davidson's Pasteurized Eggs; they look and taste like unpasteurized eggs. This new technique, allowing in-shell pasteurization, may "egg on" other producers to make pasteurized eggs more available.

There is some comfort in knowing that city and state health departments, as well as agencies at the national and international levels, including the CDC, the USDA, and the WHO, are all working to develop more efficient strategies of surveillance and of standardization of sampling procedures. In the United States the largest meat and poultry processing plants have been required by the USDA since 1998 to implement the Hazard Analysis and Critical Control Point program targeted at reducing contamination during food processing. In 1997 PulseNet, a network designed to track foodborne illnesses at a national level, was established, allowing comparison of DNA fingerprint patterns through electronic communication. About a year later PulseNet revealed that isolates of *Listeria* responsible for outbreaks of infection involving an estimated 100 cases and twenty-two deaths were linked to contaminated hot dogs and deli meats from a single processing plant.

It is convenient to rely on the food industry and governmental agencies as guardians to protect consumers from the perils of foodborne diseases. On the other hand, the specter of mad cow disease, a prion-caused disease, and the continued outbreaks of foodborne disease sound a word of caution the world over. Consumers need to share in the responsibility of minimizing the consumption of contaminated foods by exercising common sense without becoming food safety fanatics.

AUTHOR'S NOTE
About 175 years ago Jean Anthelme Brillat-Savarin, a renowned gastronome, stated, "Tell me what you eat and I will tell you what you are." What do you think of this?

■ Immunization

The 1999 CDC report, *Ten Great Public Health Achievements in the United States, 1900–1999,* includes vaccination (immunization) as one of the achievements (Table 8.1). No wonder, considering the millions of lives saved around the world as a result of vigorous vaccination campaigns targeted against bacterial and viral diseases. In the United States alone the lives of three million children are saved each year because of routine immunization. For some diseases the decline over the past century is 100%, or close to 100%, in the United States (TABLE 8.4). As examples, according to the CDC, in 1999 there were 900,000 fewer cases of measles than in 1941 and 21,000 fewer cases of polio than in 1951.

TABLE 8.4 Impact of Twentieth-Century Public Health Achievements on Selected Diseases in the United States

Disease	% Decline
Smallpox	100
Diphtheria	100
Poliomyelitis (paralytic)	100
Measles	100
Haemophilus influenzae type B	99.7
Mumps	99.6
Congenital rubella syndrome	99.4
Rubella	99.3
Tetanus	97.4
Pertussis	95.7

Adapted from CDC, *Morbidity and Mortality Weekly Report* 48 (1999):241–243.

Smallpox was the first disease to be eradicated due to Edward Jenner's pioneering work with vaccination and to the fact that the disease has no animal reservoirs. In 1881, almost a century after Jenner's work, Louis Pasteur developed a vaccine against anthrax in animals and, only a few years later (1885), developed a vaccine against human rabies. Before 1900 vaccines against three additional diseases were available, and during the twentieth century twenty-one additional diseases were added to the list. In 2006 three more vaccines were approved by the U.S. Food and Drug Administration (FDA) (TABLE 8.5). The FDA is the agency responsible for issuing a license to vaccine manufacturers in the United States, allowing the vaccine to be widely distributed. Polio is near eradication, and other diseases are on the "hot list" for eradication, including guinea worm and measles. Vaccine development is an area of intensive research. Vaccines against AIDS, tuberculosis, and malaria—diseases that kill more than five million people each year, including many children—are badly needed.

The availability of vaccination is a story of "good news and bad news." The good news is that much of the world is immunized against a variety of microbial diseases, whereas the bad news is that underdeveloped countries have not shared in this advance. Tragically, whooping cough, measles, and tetanus take the lives of over three million children each year, despite the fact that immunization is available. The Bill and Melinda Gates Foundation's Child Vaccination Program, The Global Fund for Children, The Clinton Foundation, and The Global Alliance for Vaccines and Immunization are but a few examples of organizations aimed at delivering vaccines to the world's poorest countries. Recombinant vaccines, including recombinant plant vaccines, are described below and allow easier vaccine distribution to less-developed countries.

Active Immunization

TABLE 8.6 outlines the categories of immunization. Active immunization is the result of stimulating a person's immune system to produce antibodies and memory cells

TABLE 8.5 Vaccine-Preventable Diseases

Disease	Year of Vaccine Development or U.S. Licensure
Smallpox (V)	1798
Rabies (V)	1885
Typhoid (B), cholera (B)	1896
Plague (B)	1897
Pertussis (B)	1926
Tetanus (B), tuberculosis (B)	1927
Influenza (V)	1945
Yellow fever (V)	1953
Poliomyelitis (V)	1955
Measles (V)	1963
Mumps (V)	1967
Rubella (German measles) (V)	1969
Anthrax (B)	1970
Meningitis (B)	1975
Pneumonia (B)	1977
Adenovirus (V)	1980
Hepatitis B (V)	1981
Haemophilus influenzae type b (B)	1985
Japanese encephalitis (V)	1992
Hepatitis A (V), chickenpox (V)	1995
Lyme disease (B)	1998
Pneumococcal conjugate (B)	2000
Meningococcal conjugate (B)	2005
Shingles (V), rotavirus (V), human papilloma virus (V)	2006

B, bacterial disease; V, viral disease.
Adapted and updated from CDC, *Morbidity and Mortality Weekly Report* 48 (1999):241–243.

TABLE 8.6 Outline of Immunization

Active immunization (individual makes own antibodies)
   Natural (subclinical or clinical disease and recovery)
   Artificial (vaccines for immunization "shots")
   Live, attenuated microbes
   Killed microbes
   Toxoids and other purified microbial components
   New and experimental vaccines
Passive immunization (individual receives preformed antibodies)
   Natural (in utero mother-to-infant passage; breast milk)
   Artificial (use of immune globulin)

TABLE 8.7 An Ideal Vaccine

General Vaccine Requirements	Ideal Vaccine Requirements
Safety	Safety
Effectiveness	Effectiveness
Stability	Stability
	Affordability
	Administration as a nasal spray or edible vaccine
	No need for refrigeration; stability at ordinary "tropical" temperatures
	One dose or one shot
	Long shelf life

and generally confers immunity over a relatively long time. There are two categories of active immunization—natural and artificial. Natural active immunity is achieved by the natural process of recovering from a particular disease. Analysis of an individual's blood serum frequently reveals the presence of antibodies against which there is no clinical history of disease, indicating that the disease at a subclinical level had occurred.

The use of vaccines, however, is artificial in the sense that vaccines are administered into the body to provoke an antibody immune response as a future protective measure. As indicated in Table 8.6, artificial active immunization can be accomplished in four ways; the method of active immunization reflects the best protection for the particular disease. All strategies must meet three basic requirements: safety, effectiveness, and stability (TABLE 8.7). Additionally, an ideal vaccine needs to be affordable to developing countries.

Safety issues are of prime importance and are further addressed below under Vaccine Safety. For a vaccine to be effective it must stimulate an immune response that affords protection to vaccine recipients. Vaccine preparations need to be stable over time to make them cost-effective. Some vaccines can be stored at room temperature, whereas others need to be refrigerated, presenting a problem in distribution of these vaccines to the developing world. On too many occasions vaccines need to be destroyed because refrigeration requirements are not observed. Strict attention needs to be paid to expiration dates and to conditions of storage.

Types of Active Artificial Vaccines

How are live, disease-producing microbes turned into non–disease-producing, antibody-stimulating agents? As indicated in Table 8.6, artificial active vaccines can be produced in four ways.

Live Attenuated Microbes

Vaccines made from live attenuated microbes, most of which are viruses, confer long-lasting immunity, frequently lifetime. (Poetic license is taken with the term "live" to describe viruses.) The word "attenuated" means weakened in terms of

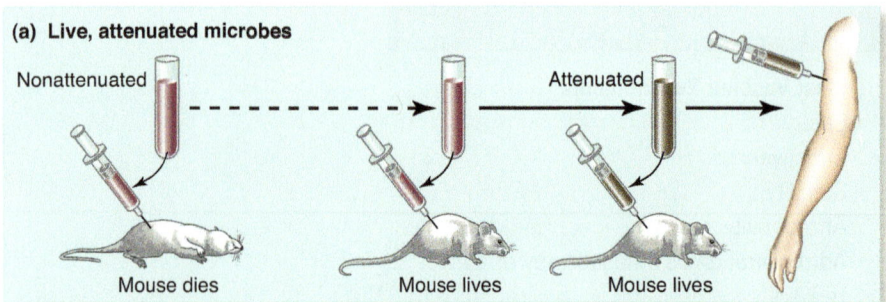

(a) Live, attenuated microbes

Nonattenuated   Attenuated

Mouse dies   Mouse lives   Mouse lives

FIGURE 8.11 Preparation of a vaccine. **(a–d)** The dashed line in panel **a** indicates that the number of successful transfers is indeterminate and that the transfers continue until attenuation takes place as determined by loss of virulence as tested in mice.

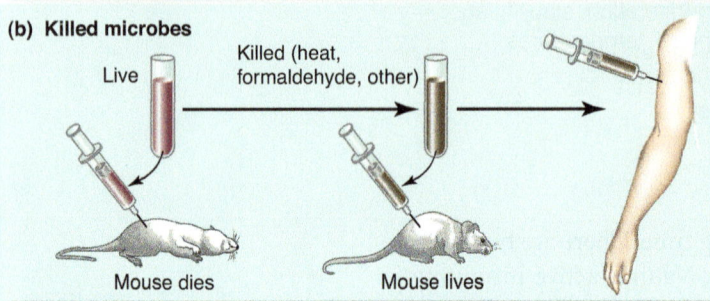

(b) Killed microbes

Live   Killed (heat, formaldehyde, other)

Mouse dies   Mouse lives

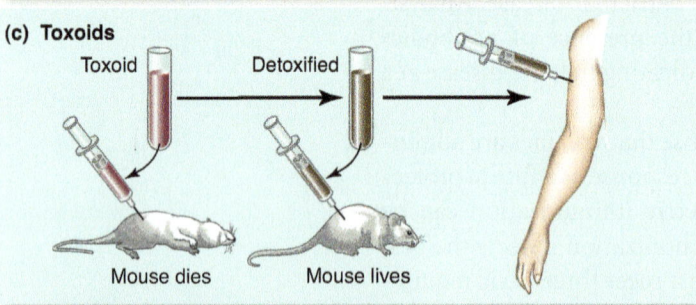

(c) Toxoids

Toxoid   Detoxified

Mouse dies   Mouse lives

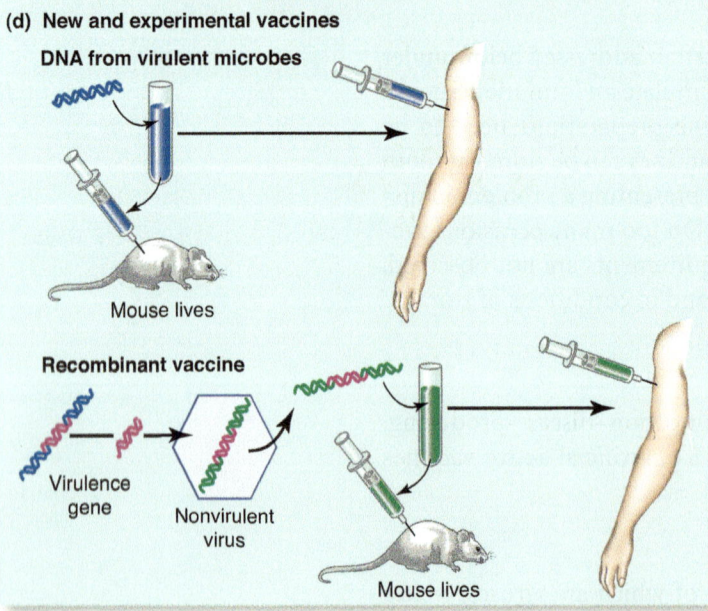

(d) New and experimental vaccines

DNA from virulent microbes

Mouse lives

Recombinant vaccine

Virulence gene   Nonvirulent virus

Mouse lives

virulence. These vaccines may produce mild and limited symptoms but not overt clinical disease, as manifested by nausea, headache, fatigue, and soreness at the site of injection for about twenty-four hours; frequently, there are no symptoms. Some live attenuated viral vaccines have been achieved by serial (repeated) transfer in tissue culture (FIGURE 8.11), allowing the production of random and unpredictable mutants. These mutants are tested in laboratory animals, and those strains producing no symptoms are selected for further trial. The **bacillus Calmette-Guérin (BCG) vaccine** against tuberculosis continues to use a tuberculosis strain (*Mycobacterium bovis* BCG) attenuated by repeated subculturing between 1908 and 1918 on laboratory medium with no reversion to virulence for over eighty years. Pasteur's vaccine against rabies was the result of serial passage in the spinal cord of dogs and rabbits, a procedure that rendered the live virus nonvirulent but continued to provoke protective antibodies against the virus. The major concern of the use of live attenuated vaccines is the possibility of reversion to virulent forms. Examples of vaccines using the live attenuated microbes are the Sabin polio, measles, and yellow fever vaccines.

Killed (Inactivated) Microbe Vaccines

Killed microbe vaccines are used when attenuation has not been accomplished or when reversion to the virulent type is considered to be too risky. Virulent microbes are heat killed or killed with particular chemical reagents. They present no risk but are not as effective at stimulating antibody production; some require multiple doses to maintain protective antibody levels. The Salk polio vaccine and the vaccines against plague, influenza, hepatitis A, and cholera are examples.

Toxoids

Some of the most serious bacterial diseases (diphtheria, tetanus, cholera, and botulism) result from the production of very potent exotoxins. These toxins can be inactivated by heat or formaldehyde, resulting in a loss of toxicity, but they retain the property of stimulating specific antibody production; these inactivated toxins are referred to as toxoids. The diphtheria-tetanus-pertussis (DTaP) vaccine, commonly administered to children at about the age of two months, contains diphtheria and tetanus toxoids. To make antitoxins, pharmaceutical companies inject toxoids into horses (sheep or goats in some cases). After a brief time the animal's immune system produces antibodies to the toxin, and the blood is taken and processed, resulting in antitoxin. The antitoxins are used as an effective immunization strategy. Diphtheria and tetanus antitoxins are examples.

New and Experimental Vaccines

These vaccines, also called recombinant DNA vaccines, are based on DNA technology. They are substitutes for "whole agent" vaccines. A "virulence gene" is inserted into a nonvirulent host microbe and that gene then is expressed and replicated in a new host. In this new environment the gene can now be safely used as a vaccine. The vaccine against hepatitis is an example of a subunit vaccine.

DNA vaccines are another new and promising approach to immunization. In this strategy microbial DNA is inserted into plasmids that are then injected directly into the host, after which the microbial DNA is expressed by the host cells as protein. Subsequently, these proteins are recognized as foreign by the host's immune system and stimulate an immune response. DNA vaccine development for a variety of microbial diseases, including influenza, tuberculosis, malaria, Lyme disease, and hepatitis C, is underway but thus far has not been effective.

FIGURE 8.12 How immunization works.

Passive Immunization

Active immunization is based on stimulating a recipient's immune system to produce antibodies and memory cells (FIGURE 8.12). In passive immunization, by contrast, the recipient receives ready-made preformed antibodies—immune serum—from human

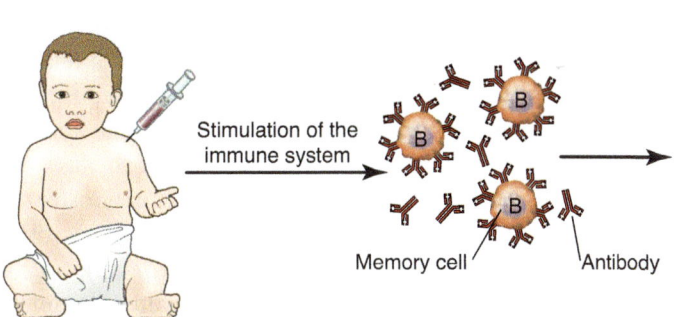

or animal sources. A big advantage to passive immunization is that antibodies are present immediately at the time of infection and can be of lifesaving value, as in cases where exposure or symptoms have already occurred. The immunity gained, however, is relatively short-lived and limited to the duration of the administered antibodies in the recipients; there is no immunological memory. TABLE 8.8 summarizes the distinctions between active and passive immunization.

Before the 1940s and the advent of antibiotics, immunotherapy—the use of immune serum—was a common practice, particularly for diphtheria, tetanus, and pneumococcal pneumonia. The first two diseases, as previously described, are toxemias; that is, the diseases are a manifestation of the production of a lethal

TABLE 8.8 Properties of Active Versus Passive Immunization

	Protection	Duration	Adverse reactions
Active	Waiting period	Extended memory	Possible
Passive	Immediate	Limited (no memory)	Possible

toxin. Antitoxins (antibodies against toxoids) are produced by the injection of toxoids (inactivated toxins) into horses to produce immune serum. In the treatment of diphtheria, about all that could be done in the days of the "horse-and-buggy" doctors was the administration of the antitoxin-containing immune serum in a desperate attempt to save lives.

Immunotherapy can be hazardous because of possible complications arising from the fact that the antibodies, along with other components of the blood serum, are "seen" as foreign protein by the recipient's immune system, resulting in antibody production against the foreign protein in the serum, causing a condition known as **serum sickness**. Serum sickness is characterized by the formation of antigen-antibody complexes that are deposited in the skin, kidney, and other body sites. Nevertheless, under certain circumstances immunotherapy is still used when immediate antibody protection is required.

The use of human immune serum, taken from individuals after vaccination or during their convalescence period from a specific disease, minimizes but does not eliminate the risk of serum sickness. This immune serum, known as human immune serum (also called immune globulins, referring to the globulin fraction of the blood), contains high levels of specific antibodies. For example, tetanus immune globulin is rich in antibodies against tetanus, and varicella-zoster immune globulin has a high concentration of antibody against the virus that causes chickenpox and shingles. Consider a case in which an individual reports to an emergency room having sustained a puncture and is at risk of tetanus. Should that person receive active immunization by a booster shot with tetanus toxoid or passive immunization with tetanus immune globulin, or both? The answer depends on the person's immune history. If the individual has, within the last ten years, received tetanus toxoid as a booster, all that is necessary is another booster to effectively stimulate those memory cells preprogrammed to produce tetanus antibodies almost immediately. However, if the individual has not received (or is not certain of) past immunizations against tetanus, immediate protection against the tetanus toxin is necessary. There is not sufficient time to make antibodies from scratch, in which case tetanus immune globulin should be administered, along with tetanus toxoid.

Antitoxins against the deadly toxins injected into the body by certain species of snakes and arachnids (spiders, scorpions, etc.) are examples of lifesaving passive immunization. If you are bitten by a poisonous snake or scorpion, you need immediate antibody protection.

Vaccine Safety

Vaccines have greatly reduced the burden of infectious diseases around the world, but as these diseases declined attention has focused on the risks associated with

AUTHOR'S NOTE

While in Guatemala I was about to stick my bare foot into my slipper without first checking it out as I had been advised to do. Suddenly a large scorpion — at least I think it was a scorpion — crawled out. It pays to follow advice! I did see one of the world's most poisonous snakes and captured it on film. That was probably the fastest close-up I have ever taken!

vaccines; how safe are the vaccines on the market? No vaccine (or other medication) is 100% safe and without risk. The better question to be asked is, "Do the benefits of the vaccine outweigh the risk?"

For example, oral polio vaccine, a live vaccine preparation, carries a risk of polio of about one in seven million people. Now that polio is nearly eradicated, it is unfortunate that unwarranted concern is sometimes paid to the one case. This is ironic in the sense that if this disease were still prevalent, little attention would be paid to the rare and unfortunate mishap that might occur. The realization is lost that seven million people minus one individual were candidates for paralytic polio. The current strategy for polio immunization is that all four doses consist of the Salk killed virus to eliminate the slight risk of vaccine-associated paralytic polio that might occur with the Sabin live attenuated oral polio vaccine.

Vaccines are constantly monitored and modified or withdrawn as circumstances dictate. For example, in 1976 Fort Dix, New Jersey was threatened by the appearance of a new and deadly strain of swine flu viruses; in response a vaccine was quickly developed, and forty-five million people were vaccinated. Unexpectedly, in some cases the vaccine triggered a debilitating and potentially fatal neuromuscular disease called **Guillain-Barré syndrome**. Rotavirus infection is a potentially fatal disease in children, and the development of the RotaShield vaccine in 1998 was heralded as a preventive measure against this disease. About one year later RotaShield was withdrawn because of a strong association with the occurrence of intussusception (twisting and obstruction of the bowel) in twenty-three infants one to two weeks after vaccination. As disappointing as this was, it is a tribute to the FDA that they responded rapidly to this unexpected circumstance resulting in removal of the vaccine from the market. Subsequently, **RotaTeq**, a replacement for RotaShield, was approved by the FDA in 2006.

The pertussis (whooping cough) component of the combined diphtheria, tetanus, and whooping cough vaccine, formerly made from whole cells, has, since 1991, been derived from a component of the microbe, resulting in fewer adverse effects. The new DPT vaccine is called **DTaP vaccine** because of the use of acellular pertussis (aP). The old DPT, no longer used in the United States, had been associated with brain damage, autism, and learning disabilities. Autism is further discussed in BOX 8.2.

Although it is true that some adverse reactions have occurred after vaccination, it is also true that vaccines may be falsely blamed because unrelated events may coincidentally occur shortly after vaccine administration. Although adverse reaction examples are worrisome, there is some comfort in the realization that the FDA attempts to stay on top of the situation.

In response to the vaccine safety concerns, Congress passed the National Childhood Vaccine Injury Act in 1986, mandating that health care providers furnish a vaccine information sheet to recipients describing the risks and benefits of the vaccine. Providers are also required to report certain side effects after vaccination to the FDA's Vaccine Adverse Event Reporting System. Further, "no-fault" vaccine compensation is provided to those injured by vaccines.

The FDA, the licensing agent, does not approve a vaccine unless initial trials indicate the benefits clearly outweigh the risks. Licensure is a rigorous process involving three phases of clinical trials and may take ten or more years. Vaccines

BOX 8.2   Vaccines, Thimerosal, and Television

A handful of parent anti-vaccine groups are convinced childhood vaccines containing thimerosal, a mercury-based substance used to prevent bacterial contamination in vaccines, and in particular, MMR and DTaP, trigger autism. Autism is now recognized as a spectrum of neurological disorders from mild to severe characterized by social, behavioral, and communication problems. Even though thimerosal has not been used in childhood vaccines (except in some flu shots) since 2001, a number of studies show there has been no decline of autism. The CDC, the American Academy of Pediatrics, and other respected organizations conclude that there is no link between autism and vaccines and continue to assure parents that vaccines are safe and life-saving.

The popular media is not helping to spread the positive message to the public, however. In early January 2008 the Immunization Action Coalition (IAC), a nonprofit organization dedicated to promoting immunization, became aware that the American Broadcasting Company (ABC) was going to televise an upcoming legal drama episode in which the lawyer sues a vaccine manufacturer on the grounds that the thimerosal in the vaccine caused a child's autism. The IAC sent the letter reproduced here to ABC expressing their disapproval. Unfortunately, the letter did not persuade ABC to pull the episode. Instead, the episode was aired with the less-than-informative disclaimer, "The preceding story is fictional and does not portray any actual persons, companies, products or events" and directed viewers to the CDC autism website. The IAC letter follows:

January 25, 2008

ABC, Inc.
500 S. Buena Vista Street
Burbank, CA 91521-4551

Dear Sir or Madam:

I was dismayed to learn that ABC is planning to debut the legal drama "Eli Stone" with a script riddled with misinformation about the safety of life-saving vaccines routinely given to infants and children. My understanding, based in part on a *New York Times* article published on January 23, is that lawyer Eli Stone sues a vaccine manufacturer on behalf of the mother of an autistic child. Stone argues that the mercury-containing preservative in a vaccine the child received caused the child's autism. At the end, the jury awards the mother 5.2 million dollars.

Scientific research conducted in the past decade has decisively and repeatedly refuted the claim that childhood exposure to thimerosal, the mercury-containing preservative used in some vaccines, is a cause of autism. Most recently, the medical records of millions of Californian children who were vaccinated in the 12 years between 1995-2007 were studied. During those years, thimerosal (other than in trace amounts) was removed from all the routinely administered childhood vaccines except the influenza vaccine. The California study determined that the incidence of newly reported cases of autism *increased* during a period when the presence of thimerosal in vaccines was significantly *decreased*. The California study, along with five major studies conducted in the United States, Denmark, the United Kingdom, and Sweden, found no association between childhood vaccination with thimerosal-containing vaccines and the development of autism.

The misinformation the Eli Stone script communicates to viewers about the safety of childhood vaccines is irresponsible and dangerous. It will cause many parents to believe that vaccines are potentially hazardous to their children's health. In some instances, healthcare providers will not be able to overcome parents' fears, and the United States could see an increase in rates of life-threatening diseases, as rates of childhood immunization decline. Such fears travel to the international community. Two years ago, Nigeria experienced outbreaks of polio because of misinformation communicated by leaders in the Islamic community about the supposed "dangers" of polio vaccine.

I urge you, as a leader in the production of "Eli Stone," to do the responsible thing. Work to block the airing of this episode. In doing so, you will protect and promote the health of children within the United States and worldwide.

Sincerely,

Deborah L. Wexler, MD
Executive Director
Immunization Action Coalition

are subject to particularly high safety standards, because, unlike other health treatments, they are given as preventives to healthy people.

Vaccines are manufactured by pharmaceutical companies; each batch of vaccine must be approved by the FDA before it can be released for use by health providers. Issues of safety, effectiveness, sterility, and purity are all evaluated by laboratory procedures, and postmarketing surveillance is conducted to identify undesirable side effects that might occur in large groups of people over long periods.

Oral vaccines and vaccines administered by nasal sprays are on the horizon and may make vaccines easier to implement in developing countries. Flu Mist, introduced in 2006, is an example. Vaccines have come a long way since the pioneering work of Jenner and Pasteur.

Childhood Immunization

The burden of infectious disease has been reduced throughout the world, most notably in the United States and in other industrialized countries, through the routine practice of childhood immunization. Examples have been cited in this chapter. FIGURE 8.13 illustrates the most current recommended childhood immunization schedule in the United States, supported by the CDC Advisory Committee on Immunization Practices, the American Academy of Pediatrics, and the American Academy of Family Physicians. Included are routine immunizations against eleven diseases and immunizations against two others (hepatitis A and influenza)

FIGURE 8.13 Childhood immunization. The CDC publishes a schedule for recommended immunizations that is approved by the Advisory Committee on Immunization Practices, the American Academy of Pediatrics, and the American Academy of Family Physicians. Data from CDC and American Academy of Pediatrics.

Vaccine	Birth	1 month	2 months	4 months	6 months	12 months	15 months	18 months	19–23 months	2–3 years	4–6 years
Hepatitis B	Hep B	Hep B	Hep B		Hep B	Hep B	Hep B	Hep B			
Rotavirus			Rota	Rota	Rota						
Diptheria, Tetanus, Pertussis			DTap	DTap	DTap	DTap	DTap				DTap
Haemophilus influenzae type b			Hib	Hib	Hib	Hib	Hib				
Pneumococcal			PCV	PCV	PCV	PCV	PCV			PPSV	
Inactivated Poliovirus			IPV	IPV	IPV	IPV	IPV				IPV
Influenza					Influenza (yearly)						
Measles, Mumps, Rubella						MMR	MMR				MMR
Varicella						Varicella	Varicella				Varicella
Hepatitis A						Hep A (2 doses)	Hep A (2 doses)			Hep A series	Hep A series
Meningococcal										MCV4	MCV4

☐ Range of recommended ages

☐ Certain high-risk groups

for selected populations. Some of the immunizations are against bacterial diseases, but most are against viral diseases; attenuated, killed, subunit, and genetically engineered vaccines are all represented. All these immunizations are recommended to be started before the age of two years.

Despite the low cost and effectiveness of immunization, thousands of children and adults have never had basic immunizations or are not up-to-date. Almost 100,000 adults die every year from influenza, pneumonia, and other vaccine-preventable diseases.

■ Antibiotics

Antibiotics can rightly be considered the single most important discovery for the treatment of diseases in the history of medicine (BOX 8.3). They serve as testimony that "nature knows best." The secretion of metabolic products by soil bacteria

---

BOX 8.3   Discovery of Penicillin

The story of the discovery of penicillin is fascinating and centers on the observations of Alexander Fleming in 1928. It should be noted that others had previously described antibacterial properties of the *Penicillium* mold, but it was Fleming who followed through on a serendipitous (chance) observation. (Serendipity is a significant factor in several important scientific discoveries. Louis Pasteur wrote, "In the field of observation, chance favors only the prepared mind." In other words, it is not just sheer luck but rather the ability to recognize the significance of unexpected.) Fleming was working with staphylococci and left a Petri dish streaked with this organism on his lab bench while he went away on a two-week vacation. On returning, he noted that the plate was contaminated with a common (*Penicillium*) mold and that the staphylococci failed to grow only in the vicinity of the mold; the mold had produced an inhibitory substance—penicillin (FIGURE 8.B1). Fleming was a modest man; he stated, "Nature created penicillin. I only found it."

Penicillin's therapeutic potential was not fully investigated until several years after its discovery, when Ernst Chain, Howard Florey, Edward Abraham, and Norman Heatley purified penicillin and successfully cured mice that had been injected with fatal doses of bacteria. Human trials were initiated and proved to be very successful. World War II triggered the large-scale production of penicillin, saving thousands of soldiers' lives. By 1944 supplies of penicillin were abundant and were released for the civilian population. In 1945 Fleming was awarded the Nobel Prize in physiology or medicine, along with Chain and Florey, who helped develop penicillin into a widely available medical product.

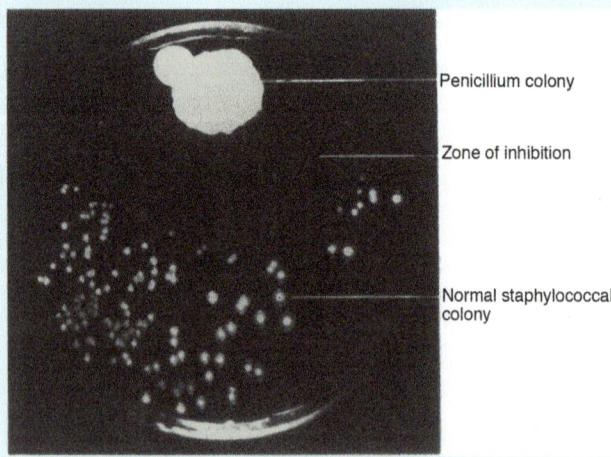

FIGURE 8.B1   Alexander Fleming's Petri dish showing the inhibition of staphylococcal colonies in the immediate vicinity of a mold contaminant, which produced penicillin. © National Library of Medicine.

and fungi that inhibit the growth of other microbes is an example of ecological antagonism at the microbial level.

History of Antibiotics

The first "wonder" drugs were the sulfonamide (sulfa) drugs, introduced in the preantibiotic era. These drugs, although they are antimicrobials, are not antibiotics, because they are synthetic compounds and not products of microbes. The sulfa drugs saved millions of lives in World War II; when sulfa drugs were not available, medics were taught to sprinkle sulfur on wounds sustained on the battlefields. An article written by Lewis Thomas, an esteemed physician, scientist, and author, appeared in the *Annals of Medicine* in 2003:

> Then came the explosive news of sulfanilamide, and the start of the real revolution in medicine. I remember the astonishment when the first cases of pneumococcal and streptococcal septicemia were treated in Boston in 1937. The phenomenon was almost beyond belief. There were moribund patients, who would surely have died without treatment, improving in their appearance within a matter of hours of being given the medicine and feeling entirely well within the next day or so.

The antibiotic era was ushered in with the first use of penicillin. On February 12, 1941 Police Constable Robert Alexander of Oxford, England was the first person in the world to receive penicillin. Alexander was seriously ill with a staphylococcal infection that started with a small sore at the corner of his mouth. Despite treatment with sulfonamides, the staphylococci spread uncontrollably into his bloodstream, resulting in numerous abscesses over his body and the spread of infection to the rest of his face, eyes, and scalp, necessitating removal of his left eye. Death seemed imminent. Miraculously, twenty-four hours after receiving penicillin he was much improved: His lesions showed signs of healing, his elevated body temperature dropped toward normal, and within several days his right eye was almost normal. Unfortunately, the small amount of penicillin that was available was insufficient for continued treatment. In a heroic effort to save the patient's life, doctors extracted penicillin from his urine and injected it back into his bloodstream. But the microbes gained the upper hand, and Alexander's condition deteriorated. He died on March 15, 1941.

After this dramatic event, antibiotics—along with improvements in sanitation and hygiene, safer foods, cleaner water, and implementation of vaccines—began contributing to the decline in infectious diseases. Antibiotics can rightly be considered the single most important discovery for the treatment of diseases in the history of medicine. They serve as testimony that "nature knows best"; the secretion of metabolic products by soil bacteria and fungi that inhibit the growth of other microbes is an example of ecological antagonism at the microbial level.

Penicillin became a prescription drug in the mid-1950s. Chemists learned how to manipulate and modify the penicillin molecule, giving rise to semisynthetic penicillin derivatives, including methicillin, ampicillin, and penicillin V, each with distinctive and beneficial properties. In the post–World War II period many other antibiotics were discovered, and their use brought about a rapid decline in deaths due to diseases caused by bacteria.

Types of Antibiotics

There is no such thing as a universal antibiotic any more than there is a universal disinfectant. Bacteria vary in their antibiotic susceptibility, and each antibiotic has a spectrum of activity against certain bacteria. Some antibiotics are more effective against gram-positive organisms, whereas others exhibit greater activity against gram-negative bacteria, but there are exceptions. A broad-spectrum antibiotic is inhibitory to a large variety of gram-positive and gram-negative bacteria, whereas a narrow-spectrum antibiotic is inhibitory to a limited range of bacteria. Some antibiotics are extremely effective but, unfortunately, exhibit marked toxicity, rendering them not useful. Some antibiotics are very expensive, whereas others are not, and some are more prone to result in antibiotic-resistant strains than others. In prescribing an antibiotic from among the many that are available, cost and antibiotic resistance are considered, but effectiveness and lack of toxicity are the central factors. A broad-spectrum antibiotic is generally prescribed when an individual is seriously ill and the causative bacteria have not been identified, so as to target a broad range of suspects. The downside of a broad-spectrum antibiotic is that a large number of bacterial species of the normal flora are killed, causing ecological disruption and allowing non–antibiotic-susceptible organisms to flourish. Some people receiving antibiotics develop oral thrush (FIGURE 8.14), a painful yeast infection, resulting from disruption of the normal bacterial flora and allowing yeasts to overgrow the normal flora. The tongue has a whitish appearance as a result of the colonies of yeast that have colonized it; fortunately, thrush responds well to mouth rinses with antiyeast drugs. In females receiving a course of antibiotics, vaginal thrush may develop and cause pain, burning, and itching, treatable by vaginal rinses. Narrow-spectrum antibiotics cause less disruption in the ecological balance of microbes and also minimize the likelihood of antibiotic resistance.

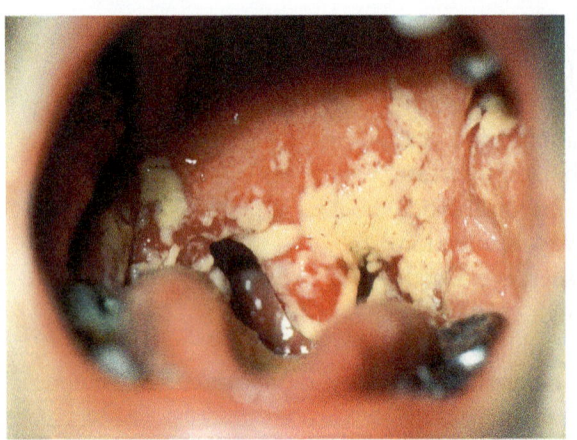

FIGURE 8.14. Oral thrush. Yellowish colored areas are growths of yeast cells. Courtesy of CDC.

Mechanisms of Antimicrobial Activity

Antibiotics act by interfering with or disrupting vital structures and metabolic pathways of the bacterial cell (FIGURE 8.15). Each antibiotic has a specific mechanism of action, although there is some overlap. For example, penicillin interferes with the ability of bacterial cells to synthesize cell walls, rendering these disabled cells subject to lysis, whereas erythromycin inhibits protein synthesis. Some antibiotics are bactericidal (they kill directly), whereas others are bacteriostatic (they keep the population from growing, thus allowing the body's defense mechanisms to get rid of the invaders). Selective toxicity is exhibited by penicillin because this drug interferes with cell wall production, and human cells do not have cell walls. On the other hand, amphotericin B is quite toxic because it acts on the bacterial cell membrane, a structure also present in human cells. The following sections are descriptions of five mechanisms of action of antibiotics:

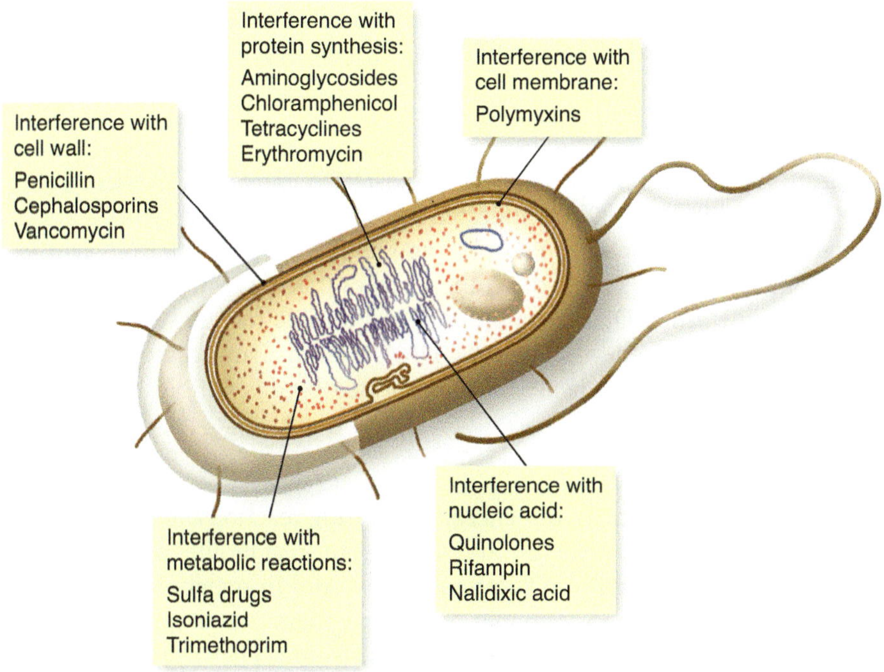

FIGURE 8.15 Mechanisms of antimicrobial activity.

Interference with protein synthesis:
Aminoglycosides
Chloramphenicol
Tetracyclines
Erythromycin

Interference with cell membrane:
Polymyxins

Interference with cell wall:
Penicillin
Cephalosporins
Vancomycin

Interference with nucleic acid:
Quinolones
Rifampin
Nalidixic acid

Interference with metabolic reactions:
Sulfa drugs
Isoniazid
Trimethoprim

Interference With Cell Wall Synthesis

Bacterial cells have rigid cell walls (peptidoglycan) that afford protection against lysis when bacteria are exposed to the low osmotic pressure of body fluids. The antibiotics penicillin and cephalosporins contain structures (beta-lactam rings) that interfere with enzymes responsible for cell wall synthesis. Vancomycin, sometimes considered the "last antibiotic stronghold," blocks a crucial reaction necessary for cell wall synthesis.

Interference With Protein Synthesis

Protein synthesis is an integral part of a cell's activity and is the culmination of expression of DNA. Bacterial ribosomes, cytoplasmic structures on which protein synthesis takes place, are targets for some antibiotics because they differ in size and structure from human ribosomes. (Bacterial cells are procaryotic and their ribosomes are 70S, whereas human cells are eucaryotic and their ribosomes are 80S. "S" represents Svedberg units, a measurement of sedimentation rates.) Streptomycin is a powerful antibiotic that was discovered in 1944; its use is now usually reserved for treatment of tuberculosis. The tetracyclines, chloramphenicol, and erythromycin are commonly used antibiotics that, like streptomycin, interfere with protein synthesis by binding with procaryotic ribosomes. Chloramphenicol is used for the treatment of typhoid fever and for other serious infections despite the fact that it can cause aplastic anemia, a potentially fatal condition in which the bone marrow ceases to produce red blood cells.

Interference With Cell Membrane Function

Cell membranes function in a vital capacity as "gatekeepers." Based on their chemical and physical structure, they control what goes into and out of the cell.

Polymyxin B is an antibiotic that binds to and distorts the bacterial cell membrane, resulting in increased permeability and leakage of important molecules out of the cell.

Interference With Nucleic Acid Synthesis

The replication and synthesis of the nucleic acids, RNA and DNA, are steps in the expression of DNA, a long and complicated series of chemical reactions that can be targeted for antibiotic activity. Rifampin and nalidixic acid block RNA synthesis. Quinolones are a large family of synthetic drugs that act by inhibiting the action of an enzyme called DNA gyrase that is responsible for the supercoiling of bacterial DNA, enabling the cell to pack DNA. Mammalian cells use different enzymes for this activity and hence are not affected by this antibiotic—another example of selective toxicity.

Interference With Metabolic Activity

Metabolism, the ability to carry out energy-generating reactions, is a key characteristic in the distinction between life and nonlife. Antimetabolites are drugs that are structurally similar to natural compounds involved in metabolism and that competitively bind with these enzymes, rendering them inactive; this is called **molecular mimicry**. The sulfa drugs work in this fashion. They mimic folic acid, a component of the microbial cell, resulting in interference with cell multiplication. Mammalian cells do not make folic acid but obtain this compound from their diet; hence, sulfa drugs can be used in human therapy.

Acquisition of Antibiotic Resistance

The experience of the past fifty years has revealed that bacteria could be thought of as "smart" because of their development of mechanisms of resistance to the antibiotics designed to bring about their death. (You really can't blame them!) But they are not really smart; the development of antibiotic resistance is a manifestation of the Darwinian process of natural selection—"survival of the fittest"—resulting from the widespread misuse of antibiotics. In Chapter 7 the development of antibiotic resistance was cited as a major international public health problem contributing to the threat of emerging infections and demands attention.

How do cells develop resistance to antibiotics? What are the biological factors involved? The development of antibiotic resistance is based on genetic changes. The bacterial cell, like all cells, has DNA and that some cells possess plasmids, extra bits of DNA independent from the DNA in the chromosome. Some plasmids are R (resistance) plasmids, which can be transferred from one cell to another; in this context they are infectious agents. Antibiotic resistance can result from mutations in the chromosomal DNA or plasmid DNA. Mutations in DNA occur spontaneously and randomly in populations of growing cells at a rate higher than one in ten million; they are not caused by selective pressure but rather are selected for survival by selective pressure. Only the survivors multiply and, in so doing, pass on their new "survival genes" along with all the other genes. The outcome is that the next generation carries these new survival genes. This is the basis for Charles Darwin's concepts of survival of the fittest and of evolution applied at the microbial level.

AUTHOR'S NOTE
You frequently hear the expression, "I'm resistant to antibiotics"; it is not you but your bacteria that display antibiotic resistance. You may be allergic, but that is a whole different story.

Human Disease and Prevention

As an analogy, consider a hypothetical population of trees in a geographical area that has suffered a drought for several years. Gradually, most of the trees die, except for a few survivors that are "lucky" enough to have random preexisting mutations in their genes that allow them to survive with less water. In the years to come the forest will be populated by drought-resistant trees. The drought did not cause the mutations but selected those trees with spontaneous, random, and preexisting mutations that allowed the trees to survive—hence, survival of the fittest. In the same fashion, antibiotics do not cause mutations but select those preexisting mutations that confer antibiotic resistance; the genes with those mutations are passed on to successive generations during cell division.

The genes for antibiotic resistance result either from chromosomal mutations or from transfer of R plasmids from antibiotic-resistant strains to antibiotic-sensitive ones. Chromosomal mutations usually confer resistance to only a single antibiotic, whereas R plasmids can confer resistance to several antibiotics at one time, a phenomenon that was first reported in Japan in 1959 when lab personnel noted the emergence of *Shigella* bacteria that were resistant to several antibiotics. The origin of these R plasmids is not known.

Transposons, or "jumping genes," are another strategy of antibiotic resistance. They may carry genes for antibiotic resistance and can integrate into chromosomes or plasmids allowing rapid dissemination of antibiotic resistance.

Mechanisms of Antibiotic Resistance

Antibiotic-resistant microbes counter the effects of the antibiotic (FIGURE 8.16) by different strategies. In some cases antibiotic resistance is the result of the production of enzymes that bring about inactivation of the antibiotic (e.g., penicillin or a cephalosporin), rendering it ineffective. Some bacteria alter the uptake of the antibiotic; they possess pumps that actively transport antibiotics out of the cell. A variety of gram-positive and gram-negative bacteria are resistant to the antibiotic

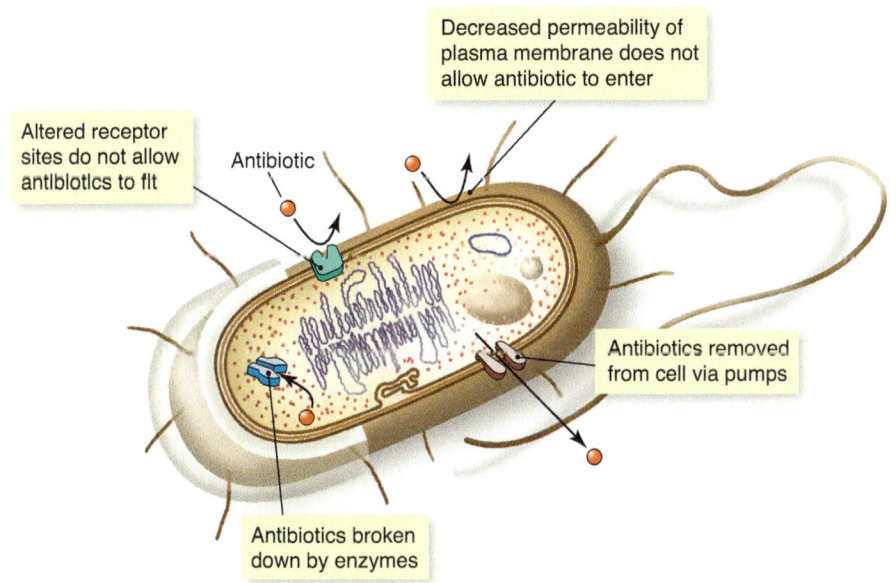

FIGURE 8.16 Microbes fight back.

Decreased permeability of plasma membrane does not allow antibiotic to enter

Altered receptor sites do not allow antibiotics to fit

Antibiotic

Antibiotics removed from cell via pumps

Antibiotics broken down by enzymes

tetracycline by this mechanism. Drug uptake may be altered by a decrease in the permeability of the cell membrane to certain antibiotics. Another mechanism of countering the effects of antibiotic activity is by modification of the drug receptor site or of an essential metabolic pathway, preventing binding of the antibiotic. Penicillin resistance in streptococci and methicillin resistance in staphylococci are the result of alteration in the antibiotic receptor sites. Finally, as pointed out above, sulfonamides and some antibiotics act by interfering with essential metabolic pathways, but some bacteria develop an alternative pathway, rendering the antibiotic ineffective.

Antibiotic resistance was not caused by the misuse of antibiotics; the genes for antibiotic resistance were present long before. Pathogenic (and other) bacteria acquired these preexisting genes from other bacteria through **horizontal gene transfer**, a mechanism by which genes are passed from one mature bacterial cell to another. (Vertical gene transfer is the passage of genes from parent to offspring.) The widespread misuse of antibiotics has fostered the emergence of antibiotic-resistant strains.

It appears that evolutionary forces are constantly at play in humans' attempts to rein in bacteria; we fight bacteria with new drugs and they fight back by adaptation. Darwin was right. Nobel Laureate Joshua Lederberg stated, "Pitted against microbial genes we have mainly our wits."

Antibiotic Misuse

Antibiotic misuse is a global problem, but the meaning of antibiotic "misuse" reveals a paradox between developed and developing countries. In developing countries antibiotics are either unavailable or affordable to the majority of the population. For those fortunate enough to procure antibiotics, it is common for people to take only a few pills and to save the rest (FIGURE 8.17) for a later time. In developed countries, where antibiotics are readily available, "misuse" is due to antibiotics being too readily available. Further, in some countries antibiotics can be purchased over the counter in markets and pharmacies without a prescription. Millions of prescriptions for antibiotics are written around the world each year, of which many are unnecessary. Prescriptions are written for colds and flu, despite the fact that antibiotics are not effective against these viral illnesses. The justification is "just in case." Too frequently, patients demand and receive antibiotics from their physicians even though antibiotics are not indicated; the particular disease is running its normal course, or the infection is caused by a virus. Studies have indicated that patients who walk out of their physician's office without a prescription for an anti-

FIGURE 8.17 The paradox of antibiotic misuse.

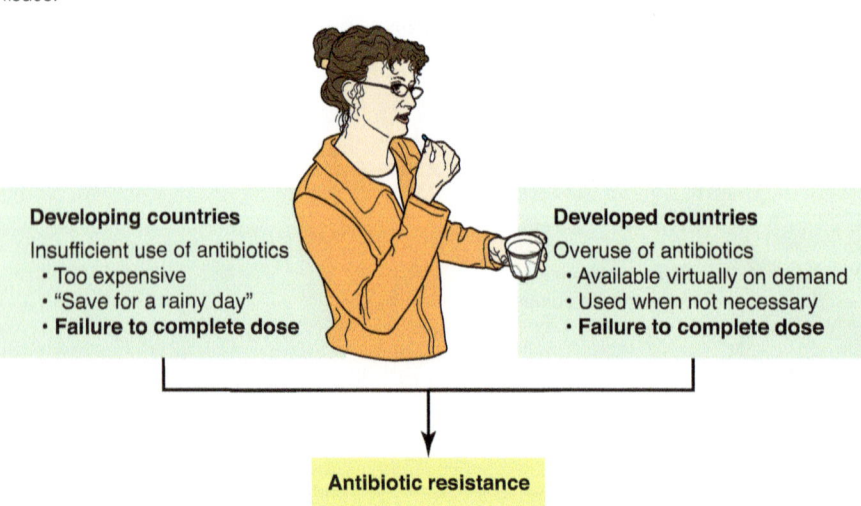

Developing countries
Insufficient use of antibiotics
• Too expensive
• "Save for a rainy day"
• **Failure to complete dose**

Developed countries
Overuse of antibiotics
• Available virtually on demand
• Used when not necessary
• **Failure to complete dose**

Antibiotic resistance

biotic will often complain that their physician "billed them for nothing." A patient's lack of knowledge about antibiotics adds to the problem. The patient begins to feel better after a few days and, because he or she is not well-informed, does not take the remaining doses. These practices favor the selection of antibiotic resistance, because only the more susceptible strains will be wiped out by only a few doses. The consequence of this misuse of antibiotics, many biologists and health professionals warn, is that we may be forced back to the preantibiotic era.

Here are a few alarming facts resulting from the misuse of antibiotics:

- Gonorrhea is increasingly more difficult to treat because of antibiotic resistance. In Hawaii, resistance to the quinolone antibiotics jumped from 1.4% in 1997 to 9.5% in 2000. The resistance of gonorrhea to a newer antibiotic, azithromycin, is on the increase.
- More than 90% of the strains of *Staphylococcus aureus* are resistant to penicillin and other antibiotics (BOX 8.4).
- Ear infections (otitis media) are the second leading cause of office visits to physicians and account for over 40% of all outpatient antimicrobial use in children. An alarming number of bacterial strains that cause this condition are now antibiotic resistant.
- Resistance to vancomycin, once considered the "last stronghold," has been reported to occur in staphylococci and enterococci.

BOX 8.4   An Urgent Public Health Problem: Methicillin-Resistant *Staphylococcus aureus*

Have you heard about methicillin-resistant *Staphylococcus aureus?* Perhaps you are more familiar with its acronym, MRSA. According to the CDC MRSA infections accounted for nearly 100,000 life-threatening illnesses and close to 19,000 deaths in 2005. It kills more Americans each year than AIDS. The appearance of MRSA infections in schools prompted these schools to close and allow cleaning personnel armed with mops and buckets of disinfectant to march in and sanitize buses, classrooms, cafeterias, and gymnasiums.

MRSA is not an infection but refers to a property of the *Staphylococcus,* namely, its resistance to the anti-staphylococcal drug methicillin and other antimicrobials, including oxacillin, penicillin and amoxicillin, and cephalosporins. Staphylococci are part of the normal flora of the skin and, for the most part, are not disease producers. *Staphylococcus aureus* is the species commonly involved in contact disease. About 50% of the population carries *S. aureus* on their skin, hair, and in their throat, and about 25% harbor this organism in their nose.

Natural selection, however, has enabled some strains of *S. aureus* to adapt to antibiotics. These strains can become dangerous when they penetrate the skin or mucous membranes and cause infections. *S. aureus* causes disease ranging from pimples and localized skin infections to life-threatening infections when it invades the blood and colonizes the internal organisms. It produces a variety of toxins that lead to serious damage.

In the past most cases were associated with hospitals and other care facilities. Disturbingly, many new cases are found in people who have no known exposure to these facilities; these cases are referred to as "community associated." Person-to-person exposure is the usual mode of transmittal through contact with infected skin lesions, nasal discharges, and contaminated hands or by contact with recently contaminated objects. The signs and symptoms of a MRSA infection include a pimple or cut that turns red and swollen, purulence (you can palpitate a fluid-filled cavity), presence of a yellow or white center or "head," draining pus, and fever.

- Drug-resistant strains of the tuberculosis bacterium are increasing worldwide.
- Approximately half of all the antibiotics produced are used for disease control and promotion of growth in animals destined for the table, a practice that fosters the selection of antibiotic-resistant pathogens in animals and a potential risk of possible transmission to humans.
- The marked increase in domestic and international travel allows exposure to antibiotic-resistant pathogens that can be spread within a country and between countries. The antibiotic-resistant strains of the gonorrhea bacterium that originated in Africa and in Asia are now prevalent throughout the world.

Working Toward the Solution

AUTHOR'S NOTE
Now that you have learned that viruses are subcellular and do not have the target sites for antibacterial activity, you know that you should not expect or ask for an antibiotic when you have a viral infection. Frequently, students complain to me about the health services on campus and complain that "the doctor didn't even give me an antibiotic!" You now know better.

The seriousness of antibiotic resistance as an impending global crisis has been established. What is the solution? The answer lies in the hands of physicians and patients, both of whom share the responsibility for the misuse and overuse of antibiotics resulting in the emergence of "superbugs," and therefore both are obligated to work toward the solution. Perhaps you unintentionally misuse antibiotics and demand that your physician prescribe an antibiotic when you are ill with what seems to be a cold. You may be guilty of not following instructions to take the full dose, because after a few days you feel better. In so doing you contribute to the emergence of antibiotic-resistant bacteria. Another too common scenario is that you do not feel well and pull out of the medicine cabinet some leftover antibiotic or ask your roommate if he or she has any antibiotics on hand without knowing the identity of the microbe. Physicians share in the responsibility for the antibiotic crisis and in its control; too often, they fail to spend the time explaining to patients why they do not need an antibiotic and succumb to patient pressure. Also, physicians fear being sued by a patient, claiming that his or her illness is a result of not being "put on an antibiotic."

Supermarket shelves are loaded with a tremendous variety of sprays, mists, and bubble-producing solutions, all designed to kill bacteria, viruses, molds and mildew. Mattresses, cribs, playpens, and toys now boast that they contain antibacterial materials. Have we gone too far? The overuse of products containing antibacterials could eventually enable the evolution of antibiotic-resistant strains, akin to the situation that exists as a result of overexposure of microbes to antibiotics. You can fight back by frequently washing your hands with regular soaps rather than hand sanitizers. The judicious use of antibiotics can stave off and minimize an already impending antibiotic-resistance crisis, allowing society to continue to enjoy the benefit of Fleming's serendipitous observation of the antagonism between a mold and a bacterium.

Antiviral Agents

There are few effective nontoxic antiviral agents. A virus infiltrates the host cell, so trying to destroy the virus while keeping damage to the cell at a minimum poses a problem. To be effective, antiviral drugs must penetrate a cell and target a stage in the viral replication cycle to block the release of new viruses. Viral replication involves the steps of adsorption, penetration, replication, assembly, and release, offering a variety of targets for effective antiviral therapy (FIGURE 8.18). As pointed out, antibiotics are not effective against viruses because they lack the target components against which antibiotics are directed. In certain circumstances, the use of antibiotics for individuals with viral infections is justified when secondary bacterial infection is a potential threat. For example, senior citizens with influenza are at risk for bacterial pneumonia, a potentially fatal disease, and are frequently put on antibiotics as a preventive measure.

A number of antiviral agents are available, and research is ongoing to develop new ones (TABLE 8.9). At some point antiviral chemotherapy may parallel antibiotic chemotherapy. The AIDS pandemic and, more recently, influenza and the potential of an avian flu outbreak drive the search for new and improved antiviral agents.

In 1999 two new antiflu drugs, zanamivir (Relenza) and oseltamivir (Tamiflu), effective against influenza A and B viruses, were approved by the FDA and are

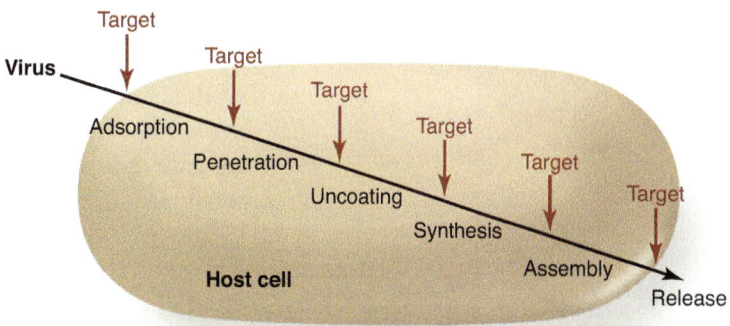

FIGURE 8.18 Mechanism of antiviral activity.

TABLE 8.9 Sampling of Antiviral Agents and Their Activity

	Mechanism of Activity
Viral entry (to synthesis)	
Amantadine	Interferes with entrance of influenza virus by blocking fusion of virus with host cell
Fuzeon	Interferes with binding of HIV to host cell receptors
Tamiflu, Relenza	Interferes with influenza neuraminidase spikes necessary for entry of virus into host cell
Viral synthesis and replication	
Acyclovir	Interferes with viral DNA replication in herpes virus
Cidofovir	Interferes with RNA, DNA synthesis; used against cytomegalovirus
Interferons	Interferes with viral replication
Nevirapine	Interferes with binding site of reverse transcriptase
Zidovudine (AZT)	Interferes with DNA replication in HIV by targeting reverse transcriptase
Viral assembly, release	
Saquinavir	Interferes with action of HV protease, resulting in noninfectious viruses

currently recommended by the CDC to treat the flu. They do not prevent or cure the flu, but if taken early they decrease duration of the illness by a few days. This may not sound like much, but if you have ever had the misfortune of having the flu, a few days is a lot!

The drug zidovudine (also called azidothymidine, or **AZT**), an inhibitor of reverse transcriptase, and a group of **protease inhibitors** have achieved some success in AIDS therapy, but, not surprisingly, drug resistance is an emerging problem. A new drug and the first integrase inhibitor, raltegravir, was approved by the FDA in October 2007 to be used with other anti-HIV agents.

■ Overview

The twentieth century witnessed an increase in life expectancy in many nations of the world. U.S. residents live thirty-three years longer, on average, than they did in 1900. At least twenty-five of those gained years are attributable to public health achievements in sanitation, clean water, food safety, immunization, and antibiotics—the topics of this chapter.

During the 1900s departments of public health designed to promote health and longevity flourished in industrialized countries of the world. Implementation of the 1972 Clean Water Act and the 1974 Safe Drinking Water Act and advances in food safety along the complex path from farm to table were major steps forward. Immunization developed from its obscure and nonscientific beginnings to result in the eradication of smallpox and the elimination, or near elimination, of a variety of infectious diseases, reducing their burden. In 2006 alone, three new vaccines became available. The availability of penicillin, followed by other antibiotics, dramatically improved the treatment of microbial diseases.

Industrialized countries of the world share in these successes. The sad part is the tremendous life span and quality of life disparity that exists between the "haves" and the "have-nots"—a highly significant world public health problem. The disparity is so great that the average life span in Macau is eighty-four years, whereas in Swaziland the life span is only thirty-two years, a fifty-two year difference. Poverty is at the root and leads to lack of clean water, inadequate sanitation, and a host of other public health deficiencies. Programs like the Kampung Improvement Program in Indonesia and the Orangi Pilot Program in Pakistan are under way, but progress is slow and disappointing.

■ Self-Evaluation

PART I: Choose the single best answer.

1. The MMR vaccine uses
   a. live viruses **b.** only live bacteria **c.** a mixture of live and dead viruses
   d. a combination of toxoids and viruses

2. The relationship between safe drinking water and child survival is
   a. inverse **b.** direct **c.** country related **d.** without correlation

3. The Kampung Improvement Program in Indonesia focused on

    a. immunization programs **b.** providing antibiotics
    c. improving the sewage system **d.** food safety

4. What agency is responsible for licensing vaccines?

    a. WHO **b.** FDA **c.** USDA **d.** CDC

5. Serum sickness can result from

    a. attenuated vaccines **b.** live vaccines **c.** toxoid administration
    d. passive immunization

6. All the following are antibiotics with the exception of

    a. chloramphenicol **b.** penicillin **c.** sulfa drugs **d.** erythromycin

PART II: Fill in the blank.

1. What is the mechanism of activity of penicillin? _____

2. Which organism is commonly used as an indicator of clean water?
_____

3. *The Jungle,* Upton Sinclair's 1906 book, has to do with _____ .

4. The DPT vaccine protects against three diseases. Name them.
_____

5. Name a microbial disease characterized by exotoxin production for which a toxoid is available for treatment. _____

6. Why is penicillin not toxic for humans? _____

PART III: Answer the following.

1. Sanitation was "in" during the twentieth century. Explain (give examples of) this statement.

2. Discuss the merits of "live versus dead" vaccines. Discuss three mechanisms by which live microbes are rendered safe for immunization purposes.

3. Describe the work of local health departments as partners in microbial disease control.

4. From a Darwinian point of view, describe the emergence of antibiotic-resistant strains.

Partnerships in the Control of Infectious Diseases

I am my brother, and my brother is me.

—Ralph Waldo Emerson

Preview

Chapter 8 emphasized the enormous strides made in the twentieth century in decreasing the burden of infectious diseases worldwide, particularly in developed countries. These successes are the result of collaborative partnerships in pooling funds, talents, and resources toward common goals. Continued progress depends on partnerships ranging from the local to the national and international levels and requires cooperation between the public and private sectors. The disparity in health care between more developed and less-developed countries requires that measures be implemented to ensure the poorer nations of the world receive their share of the benefits of progress.

Background

"The attainment for all people of the world by the year 2000 of a level of health that will permit them to lead a socially and economically productive life." This was the stated goal that emerged from the International Conference on Primary

Health Care held in Alma-Ata, USSR (now Almaty, Kazakhstan), from September 6 to 12, 1978; it was further declared:

> The Conference strongly reaffirms that health, which is a state of complete physical, mental, and social wellbeing, and not merely the absence of disease and infirmity, is a fundamental human right and that the attainment of the highest possible level of health is a most important worldwide social goal whose realization requires the action of many other social and economic sectors in addition to the health sector.

Three decades have passed since the Alma-Ata conference. Where do we stand? Has there been progress toward the attainment of these ambitious goals? Life expectancy has dramatically increased since the beginning of the twentieth century, and Alma-Ata and other international alliances have contributed to these gains. The World Health Assembly adopted the slogan "Health for all by the year 2000." That year has come and gone, and the goal has yet to be realized. Nevertheless, progress has been made, partially as a result of Alma-Ata and its emphasis on primary health care. Health for all is an elusive and moving target and may not be realistic, but striving toward it can only have positive consequences. The achievements of public health and advances in medical science have not been equally distributed across the board, resulting in poorer nations not receiving their share of the benefits of progress; this disparity remains a challenge.

The attainment of health in all its dimensions, including reducing the burden of infectious diseases, is a matter of public health concern at several levels. These levels range from the individual to the community, to state departments of health, to national agencies such as the Centers for Disease Control and Prevention (CDC) and the U.S. Food and Drug Administration (FDA), and to international agencies such as the World Health Organization (WHO). Barry Bloom, former dean of the Harvard School of Public Health, states, "One of the myths of the modern world is that health is determined largely by individual choice and is therefore a matter of individual responsibility." Realistically, individuals are limited in their efforts to achieve and maintain good health, necessitating public health strategies aimed at populations to reduce the burden of infectious and other diseases.

The eradication of infectious diseases has been a goal since the establishment of Koch's postulates and the germ theory of disease and, to date, has been achieved only in the case of smallpox. Thomas Jefferson, a few years after the introduction of smallpox vaccination in 1796, commented, "One evil more [smallpox] is withdrawn from the condition of man." In 1892 a contagious pleuropneumonia of cattle (which was imported into the United States in 1847) was declared eradicated from the country as the result of a five-year, $2 million campaign to identify and slaughter infected animals. The Rockefeller Foundation ambitiously campaigned to eradicate yellow fever and hookworm disease in the early years of the twentieth century but failed because of the complexity of eradication programs. Malaria eradication seemed plausible in the period from 1955 to 1965 but was unsuccessful, primarily because of the emergence of drug-resistant parasites and mosquitoes.

FIGURE 9.1 Smallpox: a disease of the past. The disease is characterized by the appearance of pustules on the body. The use of smallpox by terrorists is a potential worldwide threat. Courtesy of Jean Roy/CDC.

The International Task Force for Disease Eradication (ITFDE), composed of a group of scientists, convened for the tenth time since 1989 at the Carter Center in January 2007. The task force identified ninety-four diseases in terms of their potential eradication, using smallpox as the yardstick. The task force targeted three viral diseases—mumps, polio, and rubella—and three worm diseases—guinea worm disease, lymphatic filariasis, and cysticercosis—for eradication. The ITFDE defines eradication as "reduction of the worldwide incidence of a disease to zero as a result of deliberate efforts, obviating the necessity for further control measures." Elimination, according to the ITFDE, is "control of the manifestations of a disease so that the disease is no longer considered to be a public health problem," for example, blindness resulting from onchocerciasis, a worm disease, or trachoma, a chlamydial disease. Further, elimination generally refers to a limited geographical area (a single country or continent), whereas eradication is used in a global sense.

The eradication of smallpox in 1980 stands as a public health triumph of the twentieth century. Generations to come will never know the horrors of this disease, but images will serve as a reminder (FIGURE 9.1). The following factors were the unique characteristics of smallpox and of the smallpox virus that led to success in smallpox eradication and its establishment as the criterion by which to evaluate other diseases as targets for eradication:

- It is a disease only of humans; there are no natural reservoirs or biological vectors.
- The infection is easily diagnosed because of a characteristic rash.
- The duration and intensity of infectiousness is limited.
- Recovery establishes permanent immunity.
- A safe, effective, inexpensive, easily administered, stable (even in tropical climates), one-dose vaccine is available.
- Vaccination confers long-lasting immunity.
- Vaccination usually results in a permanent and recognizable scar, allowing for detection of immune versus nonimmune individuals in a population

The degree to which other diseases mimic smallpox reflects their potential for eradication, but these are not absolute criteria (BOX 9.1). For example, a biological vector is a part of the guinea worm life cycle, polio immunization requires four doses, and neither disease produces visible early manifestations. Despite these considerations, both diseases are on the "hot list" for eradication, and considerable progress has been made toward that achievement. The last case of wild polio in the Americas occurred in Peru in 1991. Not all diseases (TABLE 9.1) reviewed by the ITFDE are considered candidates for eradication for a variety of reasons, highlighting the complexity of eradication programs.

Partnerships in Infectious Disease Control

Reducing the incidence of infectious diseases is a tremendous challenge. The successes to date are largely the result of collaborative partnerships resulting in the sharing of funds, talents, and resources toward common goals. Continued successes in public health will depend on partnerships within and between the public

BOX 9.1 Criteria for Assessing Eradicability of Diseases and Conditions

Scientific Feasibility

- Epidemiological vulnerability (e.g., existence of a nonhuman reservoir, ease of spread, natural cyclical decline in prevalence, naturally induced immunity, ease of diagnosis, and duration of any relapse potential)
- Effective, safe, long-lasting, easily deployed intervention available (e.g., a vaccine or other primary preventive, a curative treatment, and a means of eliminating the vector)
- Demonstrated feasibility of elimination (e.g., documented elimination from an island or other geographical unit)

Political Will and Popular Support

- Perceived burden of the disease (e.g., extent, deaths, or other effects; true burden may not be perceived; the reverse of benefits expected to accrue from eradication; relevance to rich and poor countries)
- Expected cost of eradication (especially in relation to perceived burden from the disease)

From CDC. *Morbidity and Mortality Weekly Report* 1993;42(RR-16): 1–25.

TABLE 9.1 **International Task Force for Disease Eradication Classification**

Diseases targeted for eradication	Diphtheria (B)[a]
Dracunculiasis (W)[a]	Hookworm (W)[a]
Poliomyelitis (V)[a]	Leprosy (B)[a]
Diseases that may be eradicable	Measles (V)[a]
Lymphatic filariasis (W)[a]	Pertussis (B)[a]
Diseases of which some aspects could be eliminated	Rotaviral enteritis (V)[a]
Hepatitis B (V)[a]	Schistosomiasis (W)[a]
Neonatal tetanus (B)[a]	Tuberculosis (B)[a]
Onchocerciasis (W)[a]	Yellow fever (V)[a]
Rabies (V)[a]	Diseases not eradicable
Trachoma (B)	Amebiasis (P)[a]
Yaws (B)	Bartonellosis (B)
Diseases not eradicable now	Clonorchiasis (W)
Ascariasis (W)[a]	*Enterobius* (pinworm) disease (W)[a]
Cholera (B)[a]	American trypanosomiasis (P)[a]
	Varicella and zoster (V)[a]

[a] Discussed in this book.

B, bacterial; P, protozoan; V, virus; W, worm.

Adapted from CDC, *Morbidity and Mortality Weekly Report* 42 (1993):1–25.

---

and private sectors. Numerous agencies, both public and private, have been cited for their leadership in the fight against infectious diseases, and this chapter describes some of them.

Today's crowded societies and sharing of resources are far removed from hunter–gatherer societies, where family units were relatively isolated and depended only on their own efforts and ingenuity to stay alive. As populations grew and urbanization developed, a sharing of community responsibilities emerged in all aspects of life, including health. A negative aspect of all this "togetherness" is that the sharing of pathogens also increased. Individuals were limited in measures that could be taken to minimize exposure to pathogens. In a collective effort, the early 1900s saw the establishment of community and state departments of health that evolved into a complex network from the state level to the national and international levels and to partnerships in the public and private sectors.

At the Local Level

Community, City, and State Health Departments

Every state has a department of health, together with subordinate health departments at the community and city levels. Although the organizational charts vary from state to state, they are to a large extent a reflection of the size of the population covered. Health departments focus on the prevention of disease and the promotion of health and safety of the people within their jurisdiction. A major responsibility of health departments is to establish and implement safety regulations pertaining to food and water sanitation. For example, there is rigorous control of the milk industry in each state involving the health of dairy cows and the dairy farm conditions. Every dairy must submit milk and milk products on a strict schedule to state departments of public health laboratories to ensure the safety of the product before delivery to markets.

Public health restaurant inspectors make spot visits to eating establishments to ensure adherence to proper temperatures for the cooking and storing of foods; the absence of mice, rats, roaches, and other vermin; and the practice of appropriate measures of sanitation and hygiene by food workers. Food workers are also required to submit stool specimens to avoid a repeat of "Typhoid Mary." Salad bars must have shields to protect the food from coughs and sneezes. Increasingly, food workers must use disposable gloves (FIGURE 9.2).

FIGURE 9.2 Protection of the public. Local departments of health require food handlers to wear disposable gloves. © Sebastian Czapnik/ Dreamstime.com.

Some communities, in addition to imposing fines and closing noncompliant establishments, make restaurant inspection reports available to the public on a website and have implemented other methods to alert the public. Los Angeles requires restaurants to post inspection grades in the windows, and the local newspapers in central Florida and other areas publish the detailed results of restaurant inspections and the fines imposed on those failing to achieve a clean bill of health. In some cases the results are hard to swallow (pun intended). Violations include potentially hazardous, uncooked food held at unsafe temperatures; food handlers preparing foods with their bare hands; raw chicken stored over other raw meat in the cooler; feta cheese, ranch dressing, and cheddar cheese kept at improper temperatures; and roaches crawling over counters and food. Additionally, some communities require that food workers, in those restaurants failing to pass inspection, attend a hospitality education program.

It is not uncommon for foodborne infections to hit college campuses. Students suffer from the usual and unpleasant symptoms of gastroenteritis. Rotaviruses, noroviruses, and *Escherichia coli* are frequently identified as the causative pathogens. Each incident and its containment illustrate the epidemiological detective work necessary for tracking down the source and implementing preventive measures for the future. Investigation of such outbreaks requires partnerships at the college, community, state, and, in some cases, the national and international levels. State health departments are required to notify the CDC of specific diseases (see Chapter 4) so that a national network of surveillance and communication can be maintained (FIGURE 9.3).

Surveillance and control of infectious diseases are major functions of local health departments. For example, an outbreak of bacterial meningitis occurred in Rhode Island in 1998 and was a cause of alarm. Health officials responded with control strategies, including vaccination of thousands of schoolchildren against meningitis; the CDC was called in as a consultant. Vaccine campaigns and implementation of regulations requiring immunization against infectious diseases are another function of state health departments.

Local departments of health are involved in numerous other endeavors to foster the health and welfare of citizens, including toxicology, vital statistics, public awareness health programs, prevention and treatment of drug abuse, cancer surveillance, and lead paint screening.

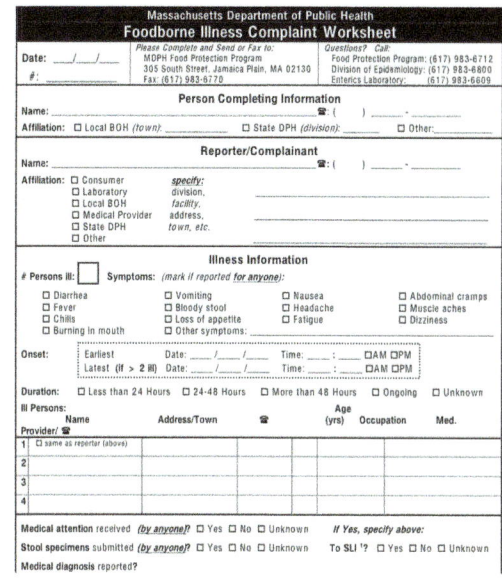

FIGURE 9.3 A reporting form that is filled out in the event of a foodborne illness and sent to a state health department. Courtesy of Massachusetts Department of Public Health.

At the National Level

Centers for Disease Control and Prevention

The CDC has been cited repeatedly throughout this book. Based in Atlanta, Georgia, it is the nation's premier public health facility, and its impact is global (FIGURE 9.4). Its functions are to

- Detect and investigate health problems
- Conduct research to enhance prevention
- Develop and advocate sound public health policies
- Implement prevention strategies
- Promote healthy behaviors
- Foster safe and healthful environments
- Provide leadership and training

The agency was founded in 1946 and employs nearly 15,000 people (including 840 commissioned corps-officers) in 170 occupations. The CDC is a vital member of partnerships with local health departments and with national and international organizations (BOX 9.2).

The CDC is the nation's main line of defense against threatening epidemics and plague. The CDC's research labs are in some ways like a large microbial zoo; within its locked

FIGURE 9.4 The CDC headquarters in Atlanta, Ga. The CDC conducts infectious disease surveillance and works in conjunction with state and local health departments and with the WHO and other agencies. Courtesy of James Gathany/CDC.

BOX 9.2   Partners in Infectious Disease Control

Local Level
  Community Departments of Health
  City Departments of Health
  State Departments of Health
National Level
  Centers for Disease Control and Prevention
  Department of Homeland Security
  U.S. Public Health Service
  Federal Emergency Management Agency
  National Institutes of Health
  American Red Cross

International Level
  World Health Organization
  Pan American Health Organization
  United Nations Foundation
  United Nations Children's Foundation
  Rotary International
  Rockefeller Foundation for the 21st Century
The Private Sector
  Pharmaceutical Companies
  The Bill and Melinda Gates Foundation
  The William J. Clinton Foundation

freezers are every known microbe on Earth (including smallpox virus) caged in small vials under the watchful eye of a microbe keeper or in the live bodies of rabbits, mice, rats, and monkeys. Down locked corridors are the "deadliest of the deadly" viruses, including those that cause HIV, rabies, Ebola hemorrhagic fever, and hantavirus pulmonary syndrome.

The CDC has come to the rescue on numerous occasions around the world helping to control or investigate incidents such as the *Legionella* outbreak in Philadelphia in 1976, the Ebola hemorrhagic fever outbreak in the Democratic Republic of the Congo in 1995 and in 2003, and the SARS epidemic in China in 2002. The CDC fields about 1,000 calls for help each year. As deemed necessary, its "SWAT" teams of epidemiologists head into the fields, frequently in collaboration with partnership agencies, equipped with ready-to-go containers stocked with syringes, needles, vaccines, intravenous fluids, examination gloves, refrigerators to store samples of blood and other tissues, generators, stacks of questionnaires, and other items to conduct guerrilla warfare against the microbes. Intriguing and heroic tales of their battles against an invisible enemy have been recounted in such movies as *Outbreak* and *Epidemic* and in books such as *The Coming Plague* and *The Hot Zone*.

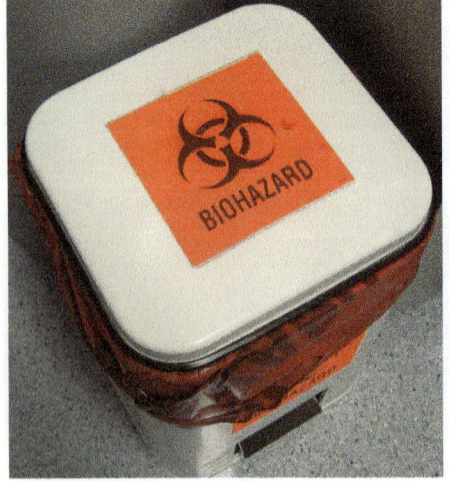

FIGURE 9.5 The biohazard symbol. Potentially infectious materials from hospitals, laboratories, and other facilities are disposed of in clearly marked biohazard containers. © joe outland/Alamy Images.

Frequently, these "disease detectives" must ship tissue samples from sick and dead victims to the CDC labs in Atlanta for identification. The labs are designed to work with deadly microbes, and their locked doors bear large biohazard signs (**FIGURE 9.5**). The CDC's Building Fifteen is the "hot zone"—a biosafety level four facility—prepared to handle the deadliest of microbes, including Ebola and hantaviruses, for which there is neither cure nor vaccine (**FIGURE 9.6**). Air leaving Building Fifteen is passed through a series of filters, water is boiled before entering sewer lines, and the fortress-like building has a camera trained on its one entrance. Researchers strip naked, shower, and don biohazard suits, known as orange suits, and as many as three pairs of gloves before entering the lab; air is supplied to them through a tube attached to the orange suit. Their protection against the deadliest of pathogens is limited to a layer of fabric, which can be penetrated by the jagged glass of a broken test tube or by a syringe needle. As much as possible, researchers work in pairs and monitor each other for fatigue and for tears in their gloves and suits.

In its role as the nation's watchdog, the CDC has developed a four-point strategy for the twenty-first century to counter the threat of new, emerging, and reemerging infections. These goals are elucidated in BOX 9.3.

Do you want to go on a cruise to some far off destination? If so, you would be well advised to consult the "green sheet," CDC's *Summary of Sanitation Inspections of International Cruise Ships*. The Vessel Sanitation Program was established in the early 1970s to minimize the risk of diarrheal diseases among passengers. All vessels with a foreign itinerary that carry more than thirteen passengers and call on U.S. ports are subject to unannounced twice-yearly inspections by Vessel Sanitation Program staff and to reinspections when necessary. To pass, the ship must score a minimum of eighty-six points on a 100-point scale. Further, the general cleanliness, personal hygiene, and physical condition of the crew, along with training programs in environmental and public health practices, are evaluated.

Which shots do you need before traveling to the Amazon, Mozambique, or Tahiti? These exotic foreign destinations are potential sources of disease that can make you very ill and even kill you. Before embarking, you can consult the CDC's website (http://www.cdc.gov) for up-to-date traveler's health information, including vaccinations, availability of safe food and water, and disease outbreaks in the land of your dreams.

FIGURE 9.6 Level four biocontainment. Ebola virus, hantaviruses, and other microbes are potentially deadly and must be handled in specially designed environments to protect laboratory personnel and the environment from the risks of contamination. Here, a lab worker in a protective suit showers inside a decontamination booth after handling pathogenic viruses. Courtesy of James Gathany/ CDC.

Department of Homeland Security

The Department of Homeland Security was proposed by President George W. Bush in June 2002 in response to the September 11, 2001 terrorist attacks on the United States. The Department replaced the earlier office of Homeland Security and entailed a major reorganization of government agencies. The Department of Homeland Security's mission, as applicable to the use of biological weapons (public health), is to

- Reduce the vulnerability of the United States to terrorism
- Minimize the damage and assist in the recovery from terrorist attacks that do occur within the United States.

U.S. Public Health Service

The Public Health Service (PHS), headed by the U.S. Surgeon General, consists of a 6,000-member Commissioned Corps and support staff and is a component of the U.S. Department of Health and Human Services (Box 9.2). It originated as the Marine Hospital Service in 1889; in 1912 the name was changed to the PHS. Initially, the PHS focused on sailors and their medical care in an attempt to alleviate the burden on public hospitals in caring for merchant seamen. In 1891 the service was charged with being the nation's medical gatekeeper by providing medical inspection of arriving immigrants to weed out "idiots, insane persons, persons likely to become a public charge, and persons suffering from a loathsome or

BOX 9.3 Preventing Emerging Infectious Diseases—CDC Strategy for the 21st Century

The CDC's strategy for combating emerging infections has four goals:

Goal I: Surveillance and Response

Objective I-A. Strengthen infectious disease surveillance and response.

Objective I-B. Improve methods for gathering and evaluating surveillance data.

Objective I-C. Ensure the use of surveillance data to improve public health practice and medical treatment.

Objective I-D. Strengthen global capacity to monitor and respond to emerging infectious diseases.

Goal II: Applied Research

Objective II-A. Develop, evaluate, and disseminate tools for identifying and understanding emerging infectious diseases.

Objective II-B. Identify the behaviors, environments, and host factors that put people at increased risk for infectious diseases and their sequelae.

Objective II-C. Conduct research to develop and evaluate prevention and control strategies in the nine target areas.

Goal III: Infrastructure and Training

Objective III-A. Enhance epidemiologic and laboratory capacity.

Objective III-B. Improve CDC's ability to communicate electronically with state and local health departments, U.S. quarantine stations, health care professionals, and others.

Objective III-C. Enhance the nation's capacity to respond to complex infectious disease threats in the United States and internationally, including outbreaks that may result from bioterrorism.

Objective III-D. Provide training opportunities in infectious disease epidemiology and diagnosis in the United States and throughout the world.

Goal IV: Prevention and Control

Objective IV-A. Implement, support, and evaluate programs for the prevention and control of emerging infectious diseases.

Objective IV-B. Develop, evaluate, and promote strategies to help health care providers and other individuals change behaviors that facilitate disease transmission.

Objective IV-C. Support and promote disease control and prevention internationally.

Adapted from *Preventing Emerging Infectious Diseases: a Strategy for the 21st Century.* CDC, 1998.

a dangerous contagious disease." The commissioner general of immunization clearly stated in 1902 that America should not become "the hospital of the nations of the earth." Immigrants were screened by PHS officers at Ellis Island for "germ diseases," including cholera, typhus, plague, smallpox, yellow fever, and trachoma (blindness). Ellis Island was dubbed "The Island of Hope" and also "The Island of Tears"—"hope" for the new and promising way of life for those who made it through the inspection line, but "tears" for those who were separated from their families and returned to their place of origin.

The PHS has taken a leading role, particularly during wars, in campaigns against sexually transmitted diseases (formerly called venereal diseases), in immunization campaigns, in vector control programs, and in public health awareness programs (FIGURE 9.7).

The mission of the PHS is to improve the health of every individual by conducting research, engineering systems for safe delivery of water and disposal

(a)

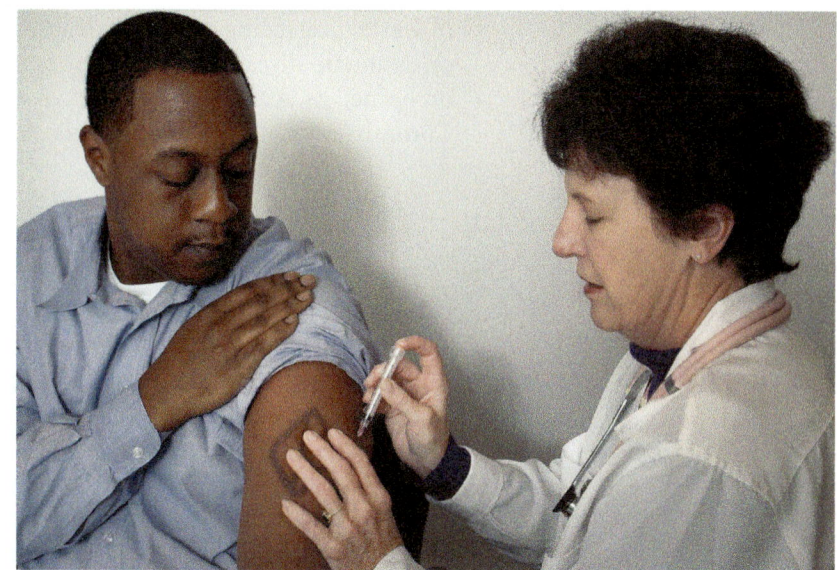

(b)

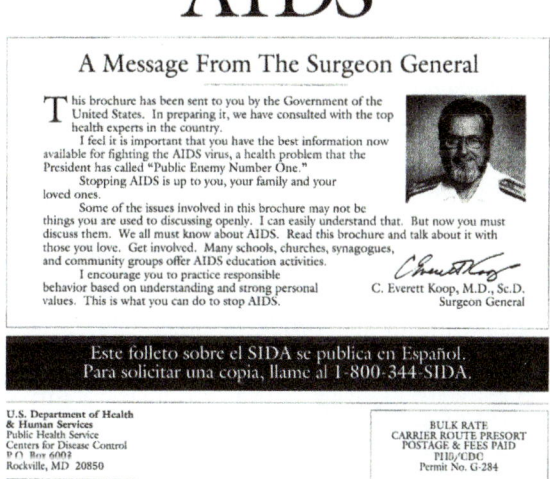

(c)

FIGURE 9.7   **(a)** The wartime fight against sexually transmitted diseases. These diseases remain a very significant public health problem in the United States and around the world. Today, AIDS overshadows the presence of other sexually transmitted diseases. Courtesy of CDC. **(b)** A person being immunized in a PHS immunization campaign. The PHS plays a major role in immunization campaigns around the world to halt epidemics by immunizing the population. Courtesy of James Gathany/CDC. **(c)** Back cover of the brochure sent to every household in the United States in the early 1990s to help Americans understand AIDS. Courtesy of CDC/U.S. Department of Health and Human Services.

of waste, overseeing food and drugs, studying and developing means to contain or eliminate disease, and promoting a safe and healthful environment at work or home. As America's uniformed service of public health professionals, the Commissioned Corps achieves this mission (according to its website) through

- Rapid and effective response to public health needs
- Leadership and excellence in public health practices
- Advancement of public health science

Federal Emergency Management Agency

The Federal Emergency Management Agency (FEMA) was created in 1979 by President Carter as an independent agency of the federal government that reports directly to the president. Its slogan is "Helping People, before, during, and after disaster." Natural disasters such as hurricanes, floods, and tornadoes can destroy the public health infrastructure, leading to polluted waters and compromised sanitation. Microbial diseases frequently follow. It is common to hear in such instances that the president has declared a community to be eligible for FEMA funds and assistance in coping with the disaster. FEMA works in partnership with local, national, and international agencies in performing its role (Box 9.2).

Hurricane Katrina presented a major challenge to FEMA in August 2005; Katrina is considered to be the largest natural disaster in the United States. The agency came under severe criticism because of a delayed response time, but, to its credit, has since implemented major policy changes and has led the way in the nation's emergency response strategies.

U.S Food and Drug Administration

The FDA is part of the PHS. It focuses on the safety of foods, vaccines, antibiotics, other medicinal (biological) products, and medical devices, all of which play a role in the prevention and control of infectious disease. This agency has the last word before these products are approved for release (see Chapter 8). The safety of the nation's blood supply system is under the umbrella of the FDA, whose inspectors routinely test blood and blood products for contamination. The agency has the authority to direct withdrawal of products, either voluntarily or legally, as in the withdrawal of the RotaShield vaccine against gastroenteritis shortly after its approval. Products not directly related to infectious diseases also fall under the scrutiny of the FDA. The agency is charged with protecting the consumer by enforcing the Federal Food, Drug, and Cosmetic Act and related public health laws, including the truth in labeling laws.

National Institutes of Health

The National Institutes of Health (NIH) is a component of the U.S. Department of Health and Human Services. The NIH is one of the most distinguished medical research centers in the world. It has come a long way since its founding in 1887 as the one-room Laboratory of Hygiene to seventy-five buildings spread over a 300-acre campus in Bethesda, Maryland (FIGURE 9.8). Composed of twenty-seven institutes

FIGURE 9.8 Building 1 at the NIH. This grant-awarding and research-oriented organization is located in Bethesda, Maryland. Courtesy of National Institutes of Health.

and centers, one of which is the National Institute of Allergy and Infectious Diseases, a primary function of the NIH is to administer and support biomedical research at over 3,000 sites in the United States and abroad; many of the research findings have direct applicability to public health measures.

At the International Level

World Health Organization

The days are long gone when a continent's, or a country's, public health problems were unique and limited to that geographical area. The planet's microbes know no boundaries and travel freely in or on their hosts without passport, from hemisphere to hemisphere in a matter of hours (see Table 7.7). The WHO headquarters in Geneva, Switzerland, functions as a command post in its extensive partnerships, and uses sophisticated systems of surveillance and communication to keep track of microbial diseases on a global level. WHO's activities are multifold and are not limited to microbial disease surveillance and control.

The WHO partnerships date back to the early years of the organization; its influenza surveillance network is responsible for identifying each year the strains of influenza to be used in the next year's vaccine, thereby serving as a global watchdog for surveillance of influenza. The WHO has entered into many partnerships in an effort to combat rabies, malaria, leprosy, sleeping sickness, filariasis, guinea worm disease, and practically every other microbial disease on the planet.

The growth of information technology provides increased opportunities for disease surveillance and response, requiring rapid assessment to initiate control efforts with minimal delay and to screen out unsubstantiated reports. In early 1997 the WHO established an innovative approach to global disease surveillance aimed at improving epidemic disease control by rapid verification and response of potentially significant outbreaks to health professionals around the globe (FIGURE 9.9). The WHO, in its alliance with 193 member countries, is in an ideal position to monitor infectious disease surveillance and control. During the verification process the WHO offers assistance in investigation and disease control, drawing on its own resources and that of its partners.

Pan American Health Organization

The Pan American Health Organization (PAHO), according to its website, has "100 years of experience in working to improve the health and living standards of the countries of the Americas." It is a component of the United Nations and serves as the Regional Office for the Americas of the WHO. Its member states include all thirty-five countries in the Americas. Its mission is achieved in association with other governmental and nongovernmental agencies, universities, and community groups.

PAHO promotes primary health strategies and assists countries in combating cholera, dengue, tuberculosis, and the spreading AIDS epidemic and is committed to ensuring that blood for transfusion is safe and not a vehicle of disease. Further, the organization aims to eliminate all vaccine-preventable diseases and embarked on polio eradication efforts in 1985, leading to a declaration of polio-free Americas in 1994; the last case was identified in August 1991. Improvement of drinking

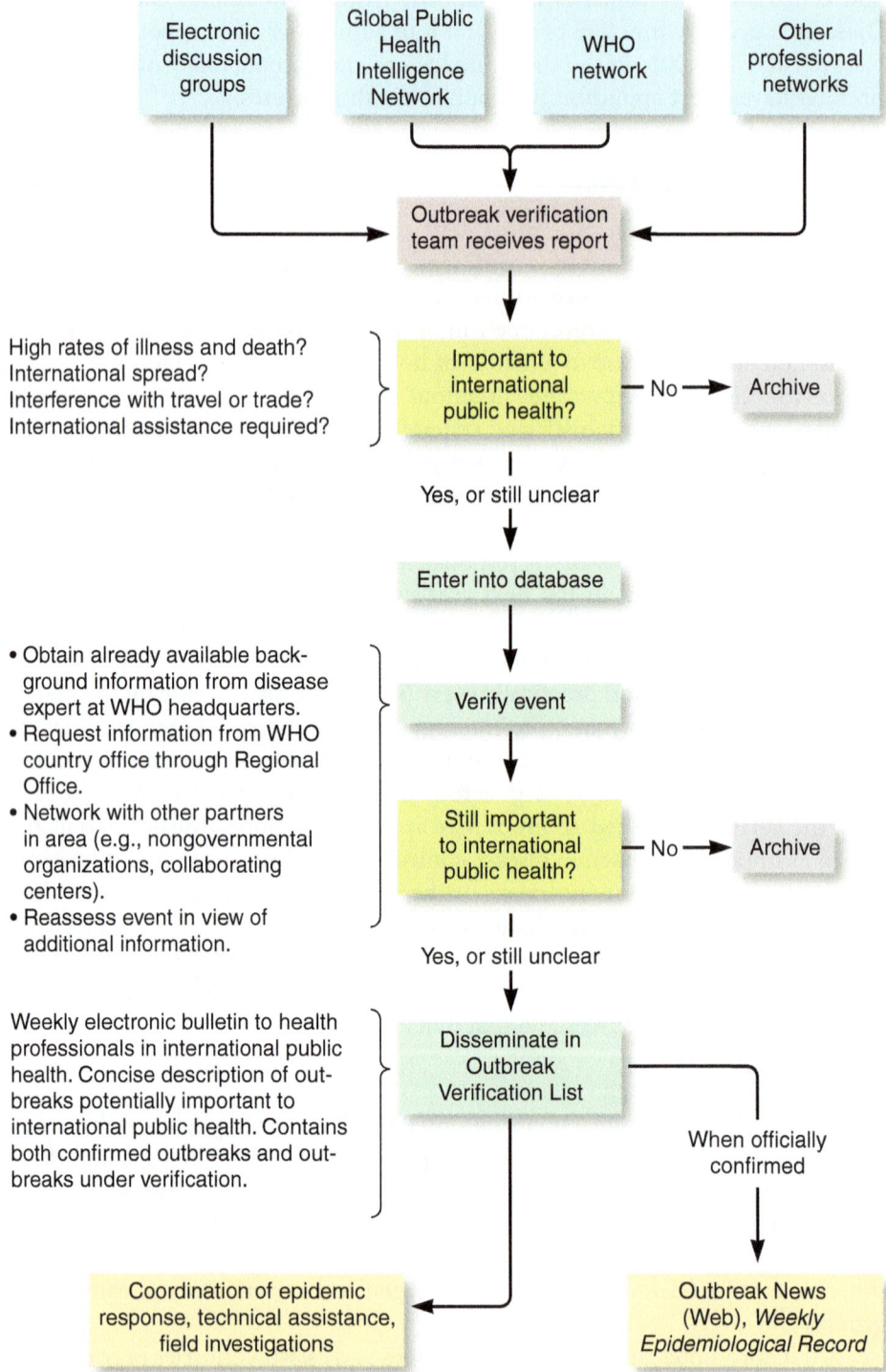

FIGURE 9.9 Outbreak verification program. This organization has teams ready to go in the event of verification of an impending outbreak of disease. Adapted from Grein, T.W., et al., *Emerging Infectious Diseases* 6 (2000): 97–102.

Electronic discussion groups

Global Public Health Intelligence Network

WHO network

Other professional networks

Outbreak verification team receives report

High rates of illness and death?
International spread?
Interference with travel or trade?
International assistance required?

Important to international public health?

No → Archive

Yes, or still unclear

Enter into database

• Obtain already available background information from disease expert at WHO headquarters.
• Request information from WHO country office through Regional Office.
• Network with other partners in area (e.g., nongovernmental organizations, collaborating centers).
• Reassess event in view of additional information.

Verify event

Still important to international public health?

No → Archive

Yes, or still unclear

Weekly electronic bulletin to health professionals in international public health. Concise description of outbreaks potentially important to international public health. Contains both confirmed outbreaks and outbreaks under verification.

Disseminate in Outbreak Verification List

When officially confirmed

Coordination of epidemic response, technical assistance, field investigations

Outbreak News (Web), *Weekly Epidemiological Record*

water, sanitation, and health care for the poor remains a top priority for PAHO, with a focus on equity.

United Nations Foundation

The UN foundation was created in 1988 with entrepreneur and philanthropist Ted Turner's historic U.S.$1 billion gift to support UN causes and activities. The UN foundation builds and implements public–private partnerships to address the world's most pressing problems and also works to broaden support for the UN through advocacy and public outreach; the UN Foundation is a public charity. The UN Foundation and its partners, including the American Red Cross and the CDC, have raised large sums of money for the Measles Initiative.

The Private Sector

The role of government agencies as partners in the control of infectious diseases has been outlined, but the success achieved over the past century is, in no small measure, also attributable to the role of the private sector and nongovernmental organizations as equal partners in the ongoing battle to lessen the world's burden of infectious diseases. Rotary International has been a key player in the polio eradication program since 1985, when it launched Polio Plus, a commitment that will cost over $500 million by the polio eradication target date of 2010. The Carter Center has committed to a variety of projects, including partnerships with both governmental and nongovernmental organizations, for the eradication of guinea worm disease, onchocerciasis, and filariasis. In January 2001 Microsoft billionaire Bill Gates pledged over $100 million to the search for an AIDS vaccine and challenged others to pitch in. Pharmaceutical companies, too, have been significant partners. Ivermectin, an important drug in the onchocerciasis control program, has been donated free of charge by Merck & Co. to countries where this disease is prevalent; Smith-Kline Beecham (now part of Glaxo Smith-Kline) has supplied the drug albendazole, an orally administered broad-spectrum antihelmintic drug, to countries as needed. Other drug companies have also made generous contributions.

The William J. Clinton Foundation was established in 1997 by former President Bill Clinton. The Clinton HIV/AIDS Initiative, created in 2002 and extended to malaria in 2007, aims to bring affordable, high-quality treatment to infected persons in developing areas.

Partnerships: The Way to Go

Chapter 8 emphasized the enormous public health strides of the last century that resulted in an increased life span for people in many countries and to decreased burden of microbial diseases around the globe. These achievements are largely the result of partnerships from the community level to the international level, as described in this chapter. The potential threat of infectious diseases is too enormous to be handled without teamwork, particularly given the explosive nature of microbial populations, the ease of transmission resulting from globalization, and the increased threat of new, emerging, and reemerging infections.

Partnerships are the key to preventing and coping with epidemics and pandemics and in responding with minimal delay to populations endangered by the ravages of infectious disease as a consequence of floods, earthquakes, and other natural disasters. History and current events provide many examples of the misery and deaths that occur in the aftermath of catastrophic events because of the collapse of the public health infrastructure. The lack of clean drinking water and safe foods, sanitation, and personal hygiene, coupled with overcrowded and makeshift living quarters, create an environment that is conducive to outbreaks of diarrheal, respiratory, and a multitude of other diseases. The circumstances are frequently more than can be handled at the local level and require the cooperation of disaster relief and public health agencies at the global level. Whereas the primary motive (one would like to believe) of these partnerships is humanitarianism, there is also a selfish motivation of self-protection, and understandably so, considering the rapid transmission of microbes from country to country.

Examples of partnerships in action in response to catastrophic events are abundant. A massive earthquake, the worst in fifty years, rumbled through western India on January 28, 2001, causing thousands of houses to collapse in Bhuj, a city of 150,000 close to the quake's epicenter, and spreading damage in its path. The earthquake was quickly followed by the arrival of international disaster teams and pledges of financial aid from Britain's International Rescue Corps, the Swiss Red Cross, the International Federation of the Red Cross, the United Nations, the European Union, Germany, Turkey, Norway, and the Netherlands. Pakistan, India's archrival, was among the first to offer condolences and demonstrated that human tragedy transcends politics.

The earthquake in India (2001), the SARS epidemic in China (2003), the tsunami in Indonesia (2004), Ebola outbreak in central Congo (2005), Hurricane Katrina in the United States (2006), and the cholera outbreak in Zimbabwe (2008) are excellent examples of partnerships in action at the local, national, and international levels, involving both public and private sectors, working together to prevent, monitor, and control infectious disease outbreaks. The world is threatened by possible pandemics of avian and swine flu, and if it were to happen, partnerships at all levels will be called on to develop and implement containment strategies. There is no doubt that everything works better when everyone works together. Partnerships are the way to go.

◼ Overview

The twentieth century witnessed tremendous accomplishments in public health that led to a worldwide decrease in the burden of infectious diseases, particularly in developed countries. The attainment of "health for all" is a goal yet to be realized. Further successes require teamwork at the local, national, and international levels and between the public and private health sectors. The players on the team are described in this chapter.

Surveillance and control of microbial disease at the community level are the responsibility of local and state departments of health. They are responsible for alerting the CDC to the occurrence of specified communicable diseases so that a national network of surveillance and communication can be maintained. At the national level a number of agencies, most notably the CDC, cooperate to bring about the containment of infectious diseases. Microbes spread readily and quickly from continent to continent, creating the need for international agencies, such as the WHO, that have the capacity to respond to outbreaks of disease at the global level.

Self-Evaluation

PART I: Choose the single best answer.

1. Early in its history, the Rockefeller Foundation was involved in attempts to eradicate

   a. smallpox **b.** hookworm **c.** AIDS **d.** malaria

2. Which characteristic of smallpox was significant in contributing to its eradication?

   a. treatable with antiviral drugs **b.** no intermediate vectors **c.** requires only two doses of vaccine **d.** can be grown on Petri dishes in the lab

3. The expression "microbial zoo" applies to

   a. WHO **b.** Bronx zoo **c.** NIH **d.** CDC

4. Which agency is the primary support of research in the United States?

   a. PAHO **b.** FDR **c.** NIH **d.** FDA

PART II: Complete the following.

1. Identify each of the letters in "ITFDE." What is the function of this organization?

2. "Hot" microbes, such as Ebola virus, are dangerous to work with and require special facilities at biosafety level _____ .

3. Rotary International continues as a key player in the campaign to eradicate _____ .

4. Name the disease used as a yardstick to evaluate potential eradication of another disease.

PART III: Answer the following.

1. Smallpox is a model chosen by ITFDE for evaluating eradication of other microbial diseases. Discuss four or five characteristics of smallpox.

2. Describe the work of local health departments as partners in microbial disease control. Cite a few specific examples.

3. You, as director of the WHO, are well aware of the disparity between the developing and developed countries in matters of public health. Outline your approach to address this issue.

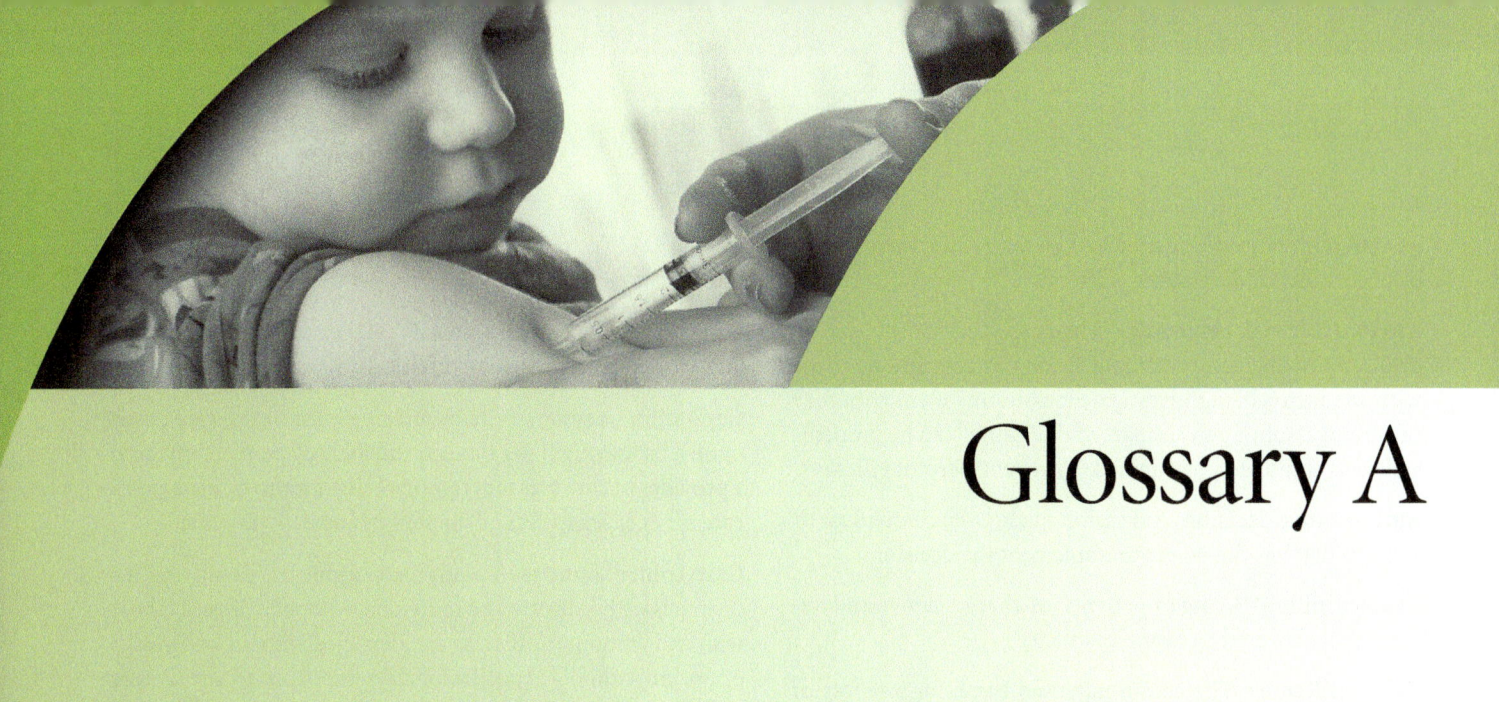

Glossary A

Absolute risk–The actual chances or probability of developing the disease expressed as a probability such as 0.01 or a percentage such as 1%.

Academic Health Center–An organization that includes a medical school, one or more other health professions schools and an affiliated hospital.

Accreditation–A process applied to educational institutions, healthcare institutions, and proposed for governmental health departments to define and enforce required structures, processes, and outcomes.

Action phase–The fourth phase of the stages of change model in which the change in behavior takes place.

Actual causes of death–Modifiable factors that lead to major causes of mortality.

Administrative law–In the United States the type of law produced by executive agencies of federal, state, and local governments.

Age-adjustment–Taking into account age-distribution of a population when comparing population or when comparing the same population at two different points in time.

Age-distribution–The number of people in each age group in a population.

All hazard approach–An approach to public health preparedness that uses the same approach to preparing for many types of disasters including use of surveillance systems, communica-tions systems, evacuations, and an organized healthcare response.

Antibody–A protein produced by the body in response to a foreign antigen which can bind to the antigen and facilitate its elimination.

Artifactual association–An association observed in the data that is actually the result of the method of data collection.

Artifactual difference or changes–Differences between pop-ulation or changes in a population over time due to changes in interest in identifying the disease, change in ability to recognize the disease or changes in the definition of the disease.

Assessment–A core public health function that includes ob-taining data that defines the health of the overall population and specific groups within the population including defining the nature of new and persisting health problems.

Association–The occurrence together of two factors, such as a risk factor and a disease, more often than expected by chance alone.

Assurance–A core public health function that includes gov-ernmental public health's oversight responsibility for ensur-ing that key components of an effective health system including health care and public health are in place even though the im-plementation will often be performed by others.

Asymptomatic–Without symptoms. When referring to screen-ing for disease, it implies the absence of symptoms of the dis-ease being sought.

At-risk population–The group of people who have a chance or probability of developing a disease.

Attributable risk percentage–The percentage of the disease or disability that can potentially be eliminated, among those with the factor being investigated, assuming a contributory cause and assuming the impact of the "cause" can be immediately and completely eliminated (*Synonym:* percent efficacy).

Authoritative decision–A decision made by an individual or a group that has the power to implement the decision.

Balance billing–Billing of patients for charges above and beyond those covered by health insurance.

Belmont Report–The commonly used name for a report of The National Commission for the Protection of Human Subjects of Biomedical and Behavioral Research which established key principles upon which the current approach to protection of human subjects is based.

Beneficence–An ethic principle which states that persons are treated in an ethical manner not only by respecting their decisions and protecting them from harm, but also by making efforts to secure their well-being.

Benefits–The positive outcomes that occur with or without an intervention.

BIG GEMS–A mneumonic which summarizes the determinants of disease including behavior, infection, genetics, geography, environment, medical care, and socioeconomic status.

Bioethics–Lies at the intersection of health law and policy and attempts to apply individual and group values and morals to controversial issues.

Biological plausibility–An ancillary or supportive criteria for contributory cause in which the disease can be explained by what is currently known about the biology of the risk factor and the disease.

Block grants–A system of federal funding to states and local jurisdictions that consolidates overall funding for a number of categories. They provide greater flexibility and allow the states to allocate funds to coordinate the delivery of services and to areas of greatest perceived need.

Branding–A marketing concepts for creating identification with a product or service that is also used in social marketing.

Built environment–The physical environment constructed by human beings.

Burden of Disease–Generically an analysis of the morbidity and mortality produced by disease. Often used to refer to the use of DALYs to estimate the burden of morbidity and mortality.

Cap–A limit on the total amount that the insurance will pay for a service per year, per benefit period, or per lifetime.

Capitation–A system of reimbursement for health care based upon a flat payment per time period for each person for whom a provider of care assumes responsibility for providing healthcare services regardless of the services actually provided.

Case-control studies–A study that begins by identifying individuals with a disease and individual without a disease. Those with and without the disease are identified without knowledge of an individual's exposure or non-exposure to the factors being investigated (*Synonym:* retrospective study).

Case-Fatality–The chances of dying from a condition once it is diagnosed.

Case finding–As used in public health an effort to identify and locate contacts of individuals diagnosed with a disease and evaluate them for possible treatment.

Categorical funding–Providing funds for public health program based upon categories such as heart disease, tuberculosis, HIV/ AIDS, maternal and child health.

Cell mediated immunity–Immunological protection that is produced by t-lymphocytes and other white blood cells that combats intracellular pathogens and tumor cells.

Certainty effect–A risk-taking attitude in which the decision maker favors the status quo rather than a probability of obtaining a better or a worse outcome.

Certification–A nongovernmental process designed to ensure competence by individual health professionals based upon completion of educational requirement and performance on an examination or other evaluation procedure.

Cohort study–An investigation that begins by identifying a group that has a factor under investigation and a similar group that does not have the factor. The outcome in each group is then assessed (*Synonym:* prospective study).

Communicable disease–A disease due to an organism such as a bacteria or virus which is transmitted person-to-person or from animals or the physical environment to human by a variety of routes from air and water, to contaminated articles or fomites, to insect bites and animal bites. Here considered a subset of infectious disease.

Community Oriented Primary Care (COPC)–A structured six step process designed to move the delivery of health services from a focus on the individual to an additional focus on the needs of communities.

Community Oriented Public Health (COPH)–An effort on the part of governmental health agencies to reach out to the community and to the healthcare delivery system to address specific health issues.

Community rating–Insurance rates set the same for all eligible individuals and families based on the previous expenses in a defined community.

Concierge practice–A form of private practice of medicine that aims to provide personalized health care to those who can afford to pay for additional access and services out-of-pocket.

Confounding variable–A difference in the groups being compared that makes a difference in the outcome being measured.

Co-insurance–The percentage of the charges that the insured is responsible for paying.

Consistency–A supportive or ancillary criteria implying that the relationship has been observed in a wide range of populations and settings.

Constitutional law–In the United States a form of law based upon the United States constitution or the constitution of a state.

Contemplation phase–The second phase of the stages of change model in which an individual is actively thinking about the benefits and barriers to change.

Contributory cause–A definition of causation that is established when all three of the following have been established: (1) the existence of an association between the "cause" and the "effect" at the individual level; (2) the "cause" precedes the effect in time; and (3) altering the "cause" alters the probability of the "effect."

Co-payment–An amount that the insured is responsible for paying even when the service is covered by the insurance.

Core public health functions–Describes governmental public health functions that cannot be delegated and remain the responsibility of governmental public health. The Institute of Medicine has defined these functions as assessment, assurance, and policy development.

Cost effective–A measure of the cost of an intervention relative to its benefit. A cost effective intervention implies that any additional benefit is considered worth the cost. Cost effective can also imply that a large cost savings is worth a small reduction in net effectiveness.

Covered service–A service for which health insurance will provide payment if the individual is otherwise eligible.

Credentialing–A general term indicating a process of verifying that an individual has the desirable or required qualifications to practice a profession.

Customary, prevailing, and reasonable–These standards are used by many insurance plans to determine the amount that will be paid to the provider of services.

Data–Facts or the representation of facts as opposed to information.

Database–A collection of data organized in such a way that a computer program can select and compile the desired pieces of data.

Decision Analysis–A method of quantitative decision making that incorporates probabilities and utilities and the timing of events into the process of comparing options and making recommendations.

Decision Maker–A generic term that can be applied to a range of individuals and organizations that make health decisions including individuals, health professions and organizations ranging from non-profits to corporations to government agencies.

Deductible–The amount that an individual or family is responsible for paying before being eligible for insurance coverage.

Demographic transition–Describes the impact of falling childhood death rates and extended life spans on the size of populations and the age distribution of populations.

Determinants–Underlying factors that ultimately bring about disease.

Dietary supplements–A category within FDA law that includes vitamins, minerals, and many herbal remedies.

Diffusion of Innovation Theory–A theory that identifies stages of dissemination and types of adopters of new technology and other changes including behavioral change.

Disability Adjusted Life-Years (DALYs)–A population health status measure that incorporates measures of death and disability and allow for measurement of the impact of categories of diseases and risks factors.

Discounting–A process in which we place greater importance on events which are expected to occur in the immediate future than on events that are expected to occur in the distant future.

Distribution of Disease–How a disease is spread out in a population often using factors such as person, place, and time.

Dose-response relationship–A relationship which is present if changes in levels of an exposure are associated with changes in frequency of the outcome in a consistent direction.

Downstream factors–Factors affecting behavior that directly involve an individual and can potentially be altered by individual interventions such as an addiction to nicotine.

Dread effect–Perception of an increase in the probability of occurrence of an event due to its ease of visualability and its feared consequences.

Dynamic model–An approach to understanding a problem or system that looks at the components over a period of time.

Early adopters–Individuals categorized by Diffusion of Innovation theory as those who seek to experiment with innovative ideas.

Early majority adopters–Individuals categorized by Diffusion of Innovation theory as opinion leaders whose social status frequently influences others to adopt the behavior.

Ecological assessment–An assessment of the impact of an alteration of the physical environment on plants and animals.

Effectiveness–An intervention has been shown to increase the positive outcomes or benefits in the population or setting in which it will be used.

Efficacy–An intervention increased positive outcomes or benefits in the population on which it is investigated.

Eligible–An individual may need to meet certain criteria to be eligible for enrollment in a health insurance plan.

Employer Mandates–Employers are required to directly or indirectly provide comprehensive insurance coverage for all their employees.

Epidemiological transition–A concept indicating the change that has been historically observed as part of social and economic development from mortality and morbidity dominated by infections to morbidity and mortality dominated by what has been called non-communicable disease or degenerative and man-made diseases (*Synonym:* public health transition).

Epidemiological treatment–Treatment of contacts of an individual with a disease even in the absence of evidence of transmission of the disease.

Epidemiologists–An investigator who studies the occurrence and control of disease or other health conditions or events in defined populations.

Essential public heath services–The ten services that have come to define the responsibilities of the combined local, state, and federal governmental public health system.

Estimation–A statistical term implying a measurement of the strength of an association or the size of a difference.

Evidence–Reliable quantitative or qualitative information or data upon which a decision can be based.

Experience rating–Health insurance rates set on the basis of a group's past history of healthcare expenses (*Synonym:* Medical underwriting).

False negative–Individuals who have a negative result on a screening test but turn out to have the disease.

False positive–Individuals who have a positive result on a screening test but turn out not to have the disease.

Fee-for-service–A system of reimbursement for health services provided based on charges for health services actually provided to patients.

Group association–Two factors such as a characteristic and a disease occur together more often than expected by chance alone in the same group or population. Does not require that the investigator have data on the characteristics of the individuals that make up the group or population (*Synonym:* ecological association).

Harm–The negative outcomes that may occur with or without an intervention (see side effect).

Hazard–A measure of the inherent capability of a substance to produce harm.

Health Adjusted Life Expectancy (HALE's)–A population health status measure that combines life expectancy with a measure of the population's overall quality of health.

Healthcare delivery system–A linkage of institutions and healthcare professional that together take on the responsibility of delivering coordinated care.

Healthcare system–A healthcare delivery system plus the financial system that pays for the delivery of health care.

Health Communications–Method for conveying, interpreting, and utilizing health information as the basis for decision making.

Health Informatics–Methods for obtaining and compiling health information.

Health Related Quality of Life (HRQOL)–A health status measure that reflect the number of days of unhealthy days due to physical plus mental impairment. HRQOL provides an overall quality of health measure but it does not incorporate the impact of death.

Health system—The healthcare system plus the public health system.

Herd immunity—Protection of an entire population from a communicable disease by obtaining individual immunity through vaccination or natural infections by a large percentage of the population (*Synonym:* population immunity).

Heuristics—Rules of thumb for decision making that often allow more rapid decision making based on a limited amount of information.

Home rule—Authority granted to local jurisdictions such as cities or countries by state constitutions or state legislative actions.

Immunization—The strengthening of the immune system to prevent or control disease through exposure to antigens or administration of antibodies.

Inactivated vaccine—Injection of a nonliving organism or antigens from an organism designed to develop antibodies to protect an individual from the disease (*Synonym:* dead vaccine).

Incidence—Rates which measure the chances of occurrence of a disease or other condition over a period of time usually one year.

Incremental cost effectiveness—A measurement of the additional cost relative to the additional net-effectiveness (see net-effectiveness).

Individual mandates—Individuals are required to purchase individual health insurance policies which include at least standardized minimum coverage.

Infant mortality rate—A population health status measure that estimates the rate of death in the first year of life.

Infectious disease—A disease caused by an organism such as a bacteria or virus. Here used to include communicable diseases as well as other infections that are not communicable.

Infectivity—The ability of a pathogen to enter and multiply in a susceptible host.

Inference—A statistical term used to imply the drawing of conclusions about a population based upon data from a sample using statistical significance testing.

Inform of decision—A decision making approach in which a clinician is merely expected to inform the patient of what is planned.

Information—As used here, the compiling or presenting of data for a range of uses.

Informed consent—A decision making approach in which a clinician is expected to provide information and obtain agreement to proceed from the patient.

Inpatient facility—A healthcare facility in the United States in which an individual may remain for more than 24 hours. Examples include hospitals and nursing homes.

Institutional Review Board (IRB)—An institution-based group that is mandated by federal regulations to review human research conducted at the institution and determine whether it meets federally defined research standards.

Interaction analysis—An approach to environmental health assessment that looks at the consequences of two or more exposures.

Interventions—The full range of strategies designed to protect health and prevent disease, disability, and death.

Judicial law—Law made by courts when applying statutory or administrative law to specific cases (*Synonym:* case law, common law).

Justice—An ethical principle based on a sense of fairness in distribution of what is deserved.

Koch's postulates—Four postulates that together definitely establish a cause and effect for a communicable disease: the organism must be shown to be present in every case of the disease; the organism must not be found in cases of other diseases; once isolated the organism must be capable of replicating the disease in an experimental animal; and the organism must be recoverable from the animal (see Modern Koch's postulates).

Late adopters—Those identified by Diffusion of Innovation theory as in need of support and encouragement to make adoption as easy as possible (*Synonym:* laggard).

Lead time bias—The situation in screening for disease in which early detection does not alter outcome but only increases the interval, between detection of the disease and occurrence of the outcome such as death.

Legislative law—In the United States the type of law that includes statutes passed by legislative bodies at the federal, state, and local levels.

Leverage points—Points or locations in a system in which interventions can have substantial impacts (*Synonym:* control points).

License–A legal document granted by a governmental authority that provide permission to engage in an activity such as the practice of a health profession.

Life Expectancy–A population health status measure that summarizes the impact of death in an entire population utilizing the probability of death at each age of life in a particular year in a particular population.

Live vaccines–Use of a living organism in a vaccine. Living organism included in vaccines are expected to be attenuated or altered to greatly reduce the chances that they will themselves produce disease (*Synonym:* attenuated vaccine).

Long shot effect–A decision making attitude in which a decision maker perceives the status quo as intolerable and is willing to take an action with only a small chance of success and a large chance of making the situation worse.

Mainstream factors–Factors affecting behavior that result from the relationship of an individual with a larger group or population such as peer pressure to smoke or the level of taxation on cigarettes.

Maintenance–The fifth phase of the stages of change model in which the new behavior become a permanent part of an individual's lifestyle.

Market justice–The philosophy that market forces should be relied upon to organize the delivery of healthcare services.

Medicaid–A federal-state program which covers groups defined as categorically needed as well as groups that may be covered at the discretion of the state including those defined as medically needy such as those in need of nursing home care.

Medical Home–A term describing a concept of primary care that includes a team approach as part of a larger healthcare system.

Medical loss ratio–The ratio of benefits payments paid to premiums collected, indicating the proportion of the premiums spent on medical services.

Medical malpractice–A body of state civil law designed to hold practitioners accountable to patients for the quality of health care.

Medicare–A federal health insurance system that covers most individuals 65 and older as well as the disabled and those with end-stage renal disease.

Medigap–A supplemental health insurance linked to Medicare designed to cover all or most of the charges that are not covered by Medicare including the 20% co-payment required for many outpatient services.

Modern Koch's postulates–A set of criteria for establishing that an organism is a contributory cause of a disease requiring evidence of an epidemiological association, isolation, and transmissions (see Koch's postulates).

Morbidity–A public health term to describe the frequency of impairments or disability produced by a disease or other condition.

Mortality–A public health term to describe the frequency of deaths produced by a disease or other condition.

Multiple risk factor reduction–Simultaneous efforts to reduce more than one risk factor.

Multiplicative interaction–A type of interaction between two or more exposures such that the overall risk when two or more exposures are present is best estimated by multiplying the relative risk of each of the exposures.

Natural experiment–A change that occurs in one particular population but not another similar population without the intervention of an investigator.

Necessary cause–If the "cause" is not present the disease or "effect" will not develop.

Negative constitution–The principle that United States constitution allows but does not require government to act to protect public health or to provide healthcare services.

Negligence law–A body of law designed to protect individuals from harm.

Net effectiveness–A measure of the benefits minus the harms of an intervention (*Synonym:* Net benefit).

No-duty principle–The principle of United States law that healthcare providers, either individuals or institutions, do not have an obligation to provide health services.

Nutritional transition–Countries frequently move from poorly balanced diets often deficient in nutrients and calories to a diet of highly processed food including fats, sugars, and salt.

Odds ratio–A measure of the strength of the relationship that is often a good approximation of the relative risk. This ratio is calculated as the odds of having the risk factor if the disease is present divided by the odds of having the risk factor if the disease is absent.

Outcome measures–Measures of quality that imply a focus on the result of health care ranging from rates of infection to readmissions with complications.

Out-of-pocket expenses–Payments for health services not covered by insurance that are the responsibility of the individual receiving the services.

Outpatient facility–A healthcare facility in the United States in which patients can remain for less than 24 hours. These facilities include the offices of clinicians, general and specialty clinics, emergency departments and a range of new types of community-based diagnostic and treatment facilities.

Passive immunization–Administration of antibodies to provide short term protection against a disease.

P.E.R.I. Process–A mnemonic which summarizes the evidence-based public health process including problem description, etiology, recommendations based upon evidence, and implementation.

Phenotypic expression–The clinical presentation of a disease which may be quite variable despite the same genetic composition or genotype.

Point of Service Plans (POS)–A type of health plan that is a modification of staff model HMOs that allow enrollees to obtain care outside the HMO but require that the patient pay for a portion of the cost of the care received.

Police powers–Authority of governmental public health based on the power of state government to pass legislation and implement actions to protect the common good.

Policy development–A core public health function that includes developing evidence-based recommendations and other analyses of options such as health policy analysis to guide implementation including efforts to educate and mobilize community partnerships to implement these policies.

Population comparisons–A type of investigation in which groups are compared without having information on the individuals within the group (*Synonym:* ecological study).

Population health approach–As used here, a term used to describe an evidence-based approach to problem solving that considers a range of possible interventions including health care, traditional public health and social interventions (*Synonyms:* ecological approach, socio-ecological approach).

Population health status measures–Quantitative summary measures of the health of a large population such as life-expectancy, HALEs, and DALYs.

Portability–The ability to continue employment-based health insurance after leaving employment usually by paying the full cost of the insurance. A federal law known as COBRA generally ensures employees of 18 months of portability.

Potential risk factors–Factors that are thought to be associated with an increased probability of disease.

Precontemplation phase–The first phase of the stages of change model in which an individual has not yet considered changing their behavior.

Prediction rule–A quantitative formulae designed to increase the ability to predict the outcome of an condition and thereby guide the use of interventions.

Preferred Provider Plans (PPOs)–An insurance system that works with a limited number of clinicians. These providers agree to a set of conditions that usually includes reduced payments and other conditions. Patients may choose to use other clinicians but they often need to more pay more out-of-pocket.

Preparation phase–The third phase of the stages of change model in which the individual is developing a plan of action.

Preponderance of the evidence–A legal term implying that a trial is decided based upon the conclusion that the evidence is more supportive of the plaintive than the defendant or visa versa.

Prevalence–A measurement of the number of individuals who have a disease at a particular point in time divided by the number of individuals who could potentially have the disease.

Primary care–Traditionally refers to the first contact providers of care who are prepared to handle the great majority of common problems for which patients seek care.

Primary intervention–An intervention that occurs before the onset of the disease.

Procedural due process–A form of due process that prohibits governments from denying individuals a right in an arbitrary or unfair way.

Process measures–Measurement of quality that focus on the procedures and formal processes that go in delivering care from procedures to ensure credentialing of health professionals to procedures to ensure timely response to complaints.

Proportion–A fraction in which the numerator is made up of observations that are also included in the denominator.

Protective factor–A factor which is associated with a reduced probability of disease.

Provider–A term used to include a wide range of health professionals that provide health services.

Proximate cause–A legal concept of causation which asks whether the injury or other event would have occurred if the negligent act had not occurred.

Public health assessment–A formal assessment that incorporates risk assessment but also includes data on the actual exposure of a population to a hazard.

Public health surveillance–Collection of health data as the basis for monitoring and understanding health problems, generating hypotheses about etiology, and evaluating the success of interventions (*Synonym:* Surveillance).

Quality Adjusted Life Year (QALY)–A measurement that asks about the number of life-years saved by an intervention rather than the number of lives.

Quarantine–The compulsory physical separation of those with a disease or at high risk of developing a disease from the rest of the population.

Randomization–As part of a randomized clinical trial, assignment of participants to study and control groups using a chance process in which the participants are assigned to a particular group with a known probability (*Synonym:* Random assignment).

Randomized clinical trial–An investigation in which individuals are assigned to study or control groups using a process of randomization. (*Synonym*: experimental study).

Range of normal–The range of values on a test that is established by performing the test on those who are believed to be free of a disease. (*Synonym:* reference interval).

Rates–Used here as a generic term to describe measurements that have a numerator and a denominator.

Recommendations–Statements based upon evidence indicating that actions such as cigarette cessation will improve outcomes such a reducing lung cancer.

Reductionist approach–An approach to problem solving that looks at each of the components of a problem one at a time.

Relative risk–A ratio of the probability of the outcome if a factor known as a risk factor is present compared to the probability of the outcome if the factor is not present.

Respect for persons–An ethical principle that incorporates two ethical convictions: first, individuals should be treated as autonomous agents, and second, that persons with diminished autonomy are entitled to protection.

Revenue sharing–A system of allocating federal funding to states and local jurisdictions according to a specific formula with few restrictions on its use.

Reverse causality–The situation in which the apparent "cause" is actually the "effect."

Rights–Protections afforded individuals on the basis of the United States constitution, a state constitution, or legislative actions.

Ring vaccination–As used in the smallpox eradication program, immediate vaccination of populations in surrounding geographic areas after identification of a case of disease.

Risk assessment–A process used in environmental health to formally assess the potential for harm due to a hazard taking into account factors such as the likelihood, timing, and duration of exposure.

Risk avoider–A decision maker who consistently favors avoiding an action even when a decision analysis utilizing probabilities, utilities, and timing argues for the action.

Risk indicator–A characteristic such as gender or age that is associated with an outcome but is not considered a contributory cause (*Synonym:* risk marker).

Risk factor–A characteristics of individuals or an exposure that increases the probability of developing a disease. It does not imply that a contributory cause has been established.

Risk taker–A decision maker who consistently favors taking an action even when a decision analysis utilizing probabilities, utilitizes, and timing argues against the decision.

Risk taking attitudes–A decision making attitude in which an individual or group consistently favors taking action or avoiding action that differ from the recommendations of a decision analysis utilizing probabilities, utilities, and the timing of events.

Safety net–The provision of services for those who cannot afford to purchase the services.

Sample–A small subset drawn from a larger group or population.

Score–In the context of evidence-based recommendations, a measurement of the quality of the evidence and a measurement of the magnitude of the impact.

Screening–As used here, testing individuals who are asymptomatic for a particular disease as part of a strategy to diagnose a disease or identify a risk factor.

Secondary care–Refers to specialty care provided by clinicians who focus on one or a small number of organ systems or on a specific type of service such as obstetrics or anesthesiology.

Secondary intervention–Early detection of disease or risk factors and intervention during an asymptomatic phase.

Sensitivity–A measurement of how well a test performs in the presence of disease.

Sequential testing–A screening strategy that uses one test followed by one or more additional tests if the first test is positive (*Synonym:* two-stage testing).

Shared decision making–A decision making approach in which a clinician is expected to directly or indirectly provide information and options for intervention to a patient and then rely on the patient to synthesize the information and make their own decision.

Side effect–A negative outcome that may occur from an intervention.

Single Payer–A single source of payment for all healthcare services usually envisioned as government at the state or federal level as the payer for all basic healthcare services.

Simultaneous testing–A screening strategy that uses two tests initially with follow-up testing if either test is positive (*Synonym:* parallel testing).

Skimming–Enrolling predominating healthy individuals into a health plan to reduce the costs to the plan.

Social justice–A philosophy that aims to provide fair treatment and a fair share of the reward of society to individuals and groups.

Social marketing–The use of marketing theory, skills, and practice to achieve social change, e.g., in health promotion.

Socio-economic status–In the United States a measurement using scales reflecting education, income, and professional status.

Specificity–A measure of how well a test performs in the absent of disease. Equals 1 minus the false negative rate.

Stages of Change model–A model of behavioral change that hypothesizes five steps in the process of behavioral change including pre-contemplation, contemplation, preparation, action, and maintenance.

Standard population–The age distribution of a population that is often used as the basis for comparison with other populations. The age distribution of the United States population in the year 2000 is generally used.

State Child Health Insurance Program (SCHIP)–A federally funded health insurance program that provides funds to the states to use to expand or facilitate the operation of Medicaid or other uses to serve the health needs of lower income children.

Static model–An approach to understanding a problem or system that looks at the components at one particular point in time.

Statistical significance test–A method used to draw conclusions or inferences about populations from data on sample(s) of the population.

Strength of the relationship–A supportive or ancillary criteria implying that a measurement of the strength of an association such as a relative risk or odds ratio is large or substantial.

Structural measures–Measure of quality of health care focused on the physical and organizational infrastructure in which care is delivered.

Substantive due process–A type of due process in which state and federal governments must justify the grounds for depriving an individual of life, liberty, and property.

Sufficient cause–If the "cause" is present the disease or "effect" will occur.

Supportive criteria–Criteria that may be used to argue for a cause and effect relationship when the definitive requirements have not been fulfilled (*Synonym:* ancillary criteria).

Surrogate outcome–A measurement of outcome that looks at short term results such as changes in laboratory tests that may not reflect longer term or clinical important outcomes.

Syndemic–A systems thinking approach that focuses attention on how health problems interact as part of larger systems.

System error–Problems resulting from deficiencies in the system for delivering health care or other services.

Systems analysis–An approach to environmental exposures as well as other public health problems that examines the interconnections between exposures or components of larger networks.

Tertiary care–A type of health care often defined in terms of the type of institution in which it is delivered, often academic or specialized health center. This type of care may also be defined in term of the type of problem that is addressed such as trauma centers, burn centers, or neonatal intensive care.

Tertiary intervention–An intervention that occurs after the initial occurrence of symptoms but before irreversible disability occurs.

True rate–A measurement that has a numerator which is a subset of the denominator and a unit of time, such as a day or a year, over which the number of events in the numerator is measured.

Uncontrollability effect–Perception of increased probability of occurance of an event due to the perceived inability of an individual to control or prevent the event from occurring.

Under-5 Mortality–A population health status measure that estimates the probability of dying during the first 5 years of life.

Undergraduate medical education–Refers to the four years of medical school leading to a MD or DO degree despite the fact that an undergraduate or bachelors degree is generally required for admission.

Unfamiliarity effect–Perception of increased probability of an event due to an individual's absence of prior experience with the event.

Universal coverage–Provision of at least basic or medically necessary health insurance for an entire population.

Upstream factors–Factors affecting behavior that are grounded in social structures and policies, such as government-sponsored programs that encourage tobacco production.

Utility scale–A scale that goes from 0 to 1 with zero reflecting immediate death and 1 reflecting full health. This scale is used to measure the value or important that an individual or a group places on a particular outcome.

Victim blaming–Placing the responsibility or blame for a bad outcome on the individual who experiences the bad outcome due to their behavior.

Viral load–A measurement of the quantity of a virus such as HIV which is present in the blood.

Glossary B

AB model   An explanation of the mode of toxin activity involving A and B toxin fragments.

Abiotic   Nonliving.

Acquired immune deficiency syndrome   *See* AIDS.

Active carriers   Individuals who have a microbial disease that can be transmitted to others.

Acute leukemia   One of two classifications of leukemia, in which the symptoms appear suddenly and progress rapidly and death occurs within a few months.

Adenoids   Tissues located in the back of the throat that are associated with the secondary immune structures; their lymphocytes play a role in protection from microbes entering through the nose and mouth.

Adenosine triphosphate   *See* ATP.

Adsorption   The first stage in the replication cycle of viruses during which viruses attach to the surface of host cells.

Aedes aegypti   A species of mosquito that is a vector for dengue fever.

Aedes albopictus   A species of mosquito that is a vector for dengue fever; these mosquitoes are commonly known as Asian tiger mosquitoes.

Aerobes   Organisms that require oxygen for their metabolic activities.

Aerosols   Suspensions of airborne particles ranging from 1 to 5 micrometers that are a means of transmission of microbes.

African trypanosomiasis   *See* Sleeping sickness.

AIDS   Acquired immune deficiency syndrome, a disease caused by the human immunodeficiency virus in which the immune system becomes severely compromised.

AIDS cocktail   A combination of drugs used to block the replication of human immunodeficiency virus, including two reverse transcriptase inhibitors and one protease inhibitor.

Airport malaria   Malaria that is the result of the survival of malaria-infected mosquitoes from other countries and is transmitted to people within the vicinity of an airport.

Albendazole   A single-dose drug used in the treatment of filariasis.

Algae   Photosynthetic organisms, some of which are unicellular and considered microbes.

Allergy   An adverse immune response to molecules associated with pollen, dust, foods, mites, antibiotics, or bee stings.

Alpha interferon   A drug used in the treatment of hepatitis C.

Amebiasis   A disease caused by the protozoan *Entamoeba histolytica;* it is prevalent in areas with poor sanitation.

Amino acid   Nitrogen-containing molecule that is the building block of protein.

Anaerobes   Organisms that do not require oxygen for their metabolic activities.

Anthrax   A potentially fatal disease caused by *Bacillus anthracis;* the organism has been used in bioterrorism and is a potential agent of biological warfare.

Antibiotics   Metabolic products of bacteria and fungi that inhibit the growth of other microbes.

Antibodies   Protein molecules produced in response to antigens; they react specifically with the antigen that triggered their production and are important defense mechanisms.

Antibody-mediated (humoral) immunity   One of two categories of specific immunity; it is the body's defense strategy for dealing with extracellular microbes.

Anticodon   A series of three nucleotides on tRNA that relate to codons on MRNA.

Antigen-presenting cells   Cells that phagocytize microbes and present antigenic components to other cells of the immune system.

Antigenic drift   A minor change in the H and N spikes of the influenza virus occurring over a period of years.

Antigenic shift   A major and abrupt antigenic change in the H or N spikes of the influenza virus resulting in a new strain of the virus.

Antigens   Components of microbes, usually protein structures, that are recognized as foreign by the immune system and are targeted for destruction.

APCs   *See* Antigen-presenting cells.

Arboviruses   Viruses that are borne by arthropods.

Archaea   One of three domains in the Woese system of classification, to which the archaebacteria are assigned.

Arthropoda   A biological phylum characterized by jointed appendages and divided into three classes; fleas, ticks, mosquitoes, and lobsters are examples.

Asexual stage   Part of the malaria parasite's life cycle that occurs in the liver and red blood cells of infected individuals in which sporozoites multiply without sexual reproduction taking place.

Asian tiger mosquito   *See Aedes albopictus.*

Assembly A stage in the life cycle of viruses in which the viral components are assembled.

ATP Adenosine triphosphate, a high-energy molecule that drives most cellular processes.

Autoimmune disease A disease in which the immune system fails to distinguish "self" from "nonself."

Autoinfection A process in which infection is perpetuated within the body; pinworm and strongyloidiasis are examples.

Autotrophs Microorganisms and plants that are capable of utilizing the energy of the sun or derive energy from the metabolism of inorganic compounds.

Azidothymidine *See* AZT.

AZT Azidothymidine, the first clinically safe and effective drug used for the treatment of AIDS; it acts as an inhibitor of the reverse transcriptase enzyme. Also called zidovudine.

B lymphocytes White blood cells that produce antibodies.

Babesiosis A tick-borne protozoan disease caused by species of *Babesia*.

Bacilli Rod-shaped bacteria.

Bacillus Calmette-Guérin vaccine A vaccine against tuberculosis with limited effectiveness.

Bacteria Unicellular microbes with distinct properties; one of the six distinct types of microbes.

Bacteria One of three domains in the Woese system of classification.

Basal bodies The structures by which flagella are anchored to the cell wall and cell membrane.

Basophils White blood cells that are rich in granules of histamine.

BCG *See* Bacillus Calmette-Guérin vaccine.

Beaver fever *See* Giardiasis.

Binary fission An asexual mode of reproduction in which a cell splits into two new cells.

Binomial system A system of nomenclature established by Carolus Linnaeus in 1735 in which the genus and species are identified.

Bioaugmentation Spraying of nutrients on beaches or on other microbe-contaminated sites to foster the growth of microbes indigenous to the area in order to accelerate degradation of pollutants.

Biocrime A bioterrorist act that targets a specific individual or group rather than the masses.

Biogeochemical cycles Processes such as the carbon and nitrogen cycles in which bacteria play a critical role.

Biological amplification The mechanism of activity of the complement system that promotes immune system function.

Biological vectors Organisms, including mosquitoes, ticks, lice, and flies, that transmit microbial disease.

Biological warfare The use of microbes, such as the anthrax bacillus and smallpox virus, as weapons in the conduct of war.

Biological weapons Microbes that are deployed with the intention of producing clinical disease in an attempt to incapacitate or kill large numbers of individuals.

Bioremediation A process utilizing the enzymatic activity of microbes to break down pollutants such as oil, paper, and concrete.

Biosphere All the organisms and environments found on Earth.

Bioterrorism Employment of biological weapons by nonstate governments, religious cults, militants, or crazed individuals.

Biotic Of or relating to life; refers to the living components of an ecosystem.

Biting plates Mouth parts found in some worms that allow them to attach to and suck blood from their host.

Black Death *See* Plague.

Blender An organism that facilitates the creation of a new hybrid strain of a disease; an example is that of a pig that becomes infected with a human strain and a bird strain of the flu, resulting in a new hybrid form of the virus.

Blood flukes Common name for *Schistosoma* species and some other worms.

Boils Localized skin infections frequently caused by staphylococci.

Bone marrow Source of all blood cells.

Botox Minute doses of botulinum toxin used to reduce wrinkles and to treat common disorders associated with muscle overactivity.

Botulism A form of food poisoning caused by *Clostridium botulinum;* a neurotoxin causes flaccid paralysis.

Bovine spongiform encephalopathy A neurological condition in cattle resulting from abnormally folded prions that convert normal prions to the defective form.

Breakbone fever *See* Dengue fever.

Breath test A means of diagnosing peptic ulcers based on the production of urease by *Helicobacter pylori*.

Broad-spectrum antibiotic An antibiotic that is inhibitory to a wide range of bacteria.

Bronchitis Inflammation of the bronchioles.

BSE *See* Bovine spongiform encephalopathy.

Buboes Enlarged lymph nodes that occur when bacteria localize in the lymph nodes, such as in bubonic plague.

Bubonic plague A form of *Yersinia pestis* infection in which the bacteria localize in lymph nodes, causing them to swell to the size of hens' eggs.

Budding A mechanism by which some viruses are released from the host cell.

Bulbar poliomyelitis An extremely serious form of polio in which individuals have difficulty swallowing and breathing because of muscle paralysis.

Burkitt's lymphoma A malignant cancer of the jaw and abdomen that occurs often in children in central and western Africa.

Campylobacter A bacterium that is an important cause of intestinal disease in humans and is associated with abortion and enteritis in sheep and cattle.

CA-MRSA Community associated methicillin-resistant *Staphylococcus aureus*.

Capsid   The protein coat surrounding viruses.

Capsomeres   Protein units that make up the capsid; they confer helical, polyhedral, or complex shapes.

Capsule   A component of the bacterial envelope that may contribute to virulence; it is not present in all species of bacteria.

Carbohydrates   Organic molecules that function as an energy source and are found in some cellular structures.

Carbuncles   Localized skin infections that are larger and deeper than boils and can reach baseball size.

Cauterization   An early method of treating wounds inflicted by rabid animals in which long, sharp, hot needles were inserted deeply into the wounds.

CD   *See* Cluster-of-differentiation molecules.

CD4 receptor molecules   Receptor molecules found on some T cells that identify them as T-helper cells.

CD8 receptor molecules   Receptor molecules found on some T cells that identify them as cytotoxic T cells.

Cell   The basic unit of life.

Cell culture   A technique for growing cells that can be used to culture viruses.

Cell-mediated immunity   A type of specific immunity that results in the production of sensitized T lymphocytes directed against a particular antigen.

Cell membrane   A component of the bacterial envelope that regulates the passage of molecules between the bacterial cell and its external environment.

Cell wall   A component of the bacterial envelope that gives the cell its characteristic shape and confers structural integrity.

Cercariae   Freely swimming immature forms of *Schistosoma* that are responsible for "swimmer's itch."

Cercarial dermatitis   A condition characterized by itching and a rash caused by penetration of the skin by cercariae.

Cerebral malaria   A deadly form of malaria that affects the brain.

Cervix   The opening to the uterus.

Chagas disease   A protozoan disease caused by *Trypanosoma cruzi* that leads to widespread tissue damage, particularly to the heart, causing it to enlarge and impairing its function.

Chancres   Sores on the penis or on the cervix that are characteristic of the primary stage of syphilis.

Chemosynthetic autotrophs   Organisms that derive energy from the metabolism of inorganic compounds.

Chemotaxis   The process of moving toward or away from a chemical stimulus.

Chicken pox   A disease characterized by blisterlike lesions on the body and typically occurring in childhood.

Chlamydiae   Coccoid bacteria that are obligate intracellular parasites; they are responsible for a variety of diseases, including urethritis, trachoma, and chlamydia.

Chlorella   A photosynthetic alga found on the surface of ocean water.

Cholera   A disease caused by *Vibrio cholerae* that is characterized by severe diarrhea and dehydration.

Chromosome   The structure into which DNA is organized.

Chronic carriers   Individuals who harbor a pathogen for long periods without becoming ill with the disease but who may spread the disease to others.

Chronic fatigue syndrome   A disease that was incorrectly believed to have been caused by Epstein-Barr virus.

Chronic granulomatous disease   An inherited disorder of phagocytes characterized by their inability to kill bacteria.

Chronic leukemia   A form of leukemia, cancer of the white blood cells, that can remain for many months.

Cilia   Hairlike projections that line certain areas of the respiratory system and assist in the removal of bacteria.

Ciliata   A group of protozoans characterized by the presence of cilia.

Clonal selection theory   An explanation of antibody production: A population of antibody-producing B cells exists for every possible antigen and is triggered to clone and produce antibodies upon contact with a specific antigen.

Cluster-of-differentiation molecules   Distinctive molecules found on T cells that identify their specific role.

Coagulase   An enzyme that forms a network of fibers around bacteria, affording protection against phagocytosis.

Cocci   Spherical bacteria.

Codon   A series of three nucleotides on MRNA that relates to anticodons.

Collagen   A protein found in connective tissues.

Collagenase   An enzyme that breaks down collagen.

Colonies   Visible masses of bacterial cells growing on an agar surface, each presumably derived from a single cell.

Colonization   Growth and reproduction of a microbe in a particular niche, resulting in large numbers of cells.

Colorado tick fever   A tick-borne viral fever caused by an arbovirus.

Commensalism   A symbiotic relationship between two species in which one benefits and the other is neither harmed nor benefited.

Common cold   A mild illness caused by a variety of viral groups.

Common-source epidemics   Outbreaks of disease arising from contact with a single contaminated source, typically associated with fecally contaminated food or water.

Common warts   Benign, painless, elevated growths caused by human papillomaviruses that occur most frequently on the fingers.

Competent   Cells that are able to take up DNA from the environment resulting in gene transformation.

Complement   A series of blood proteins that constitute a significant defense mechanism against disease-causing microbes.

Complex   An arrangement of capsomeres in some viruses.

Congenital syphilis   Syphilis resulting from the passage of pirochetes across the placenta from mother to baby.

Conjugation   A recombinational process resulting in the transfer of DNA from donor to recipient during physical contact.

Constitutive enzymes   Enzymes that are constantly produced because the gene switch is always in the on position.

Consumers   Organisms that take in oxygen and release carbon dioxide.

Convalescence stage   The time in which recovery from an illness takes place, strength is regained, repair of damaged tissue occurs, and rashes disappear.

Copepods   Water fleas that can harbor infective larvae of guinea worms.

Coronaviruses   A group of viruses that are a major cause of the common cold.

CPE   *See* Cytopathic effect.

Creutzfeldt-Jakob disease   A human transmissible spongiform encephalopathy.

Cryptosporidiosis   A protozoan disease transmitted by drinking fecally contaminated water.

Culture   Growth of an organism in a laboratory for the purposes of propagation and study.

Cutaneous anthrax   A form of anthrax that is acquired by contact with *Bacillus anthracis* or its spores via wool, hides, leather, or hair products.

Cutaneous leishmaniasis   A form of leishmaniasis that results in skin lesions.

Cyanobacteria   A group of bacteria that is photosynthetic.

Cyclosporiasis   A protozoan disease.

Cyst   A small sac, frequently filled with pus.

Cysticerci   Tapeworm larvae that are enclosed within a membranous sac.

Cysticercosis   A disease resulting from the ingestion of pork tapeworm eggs.

Cytokines   Products that are released by lymphocytes in response to stimuli and that trigger responses in other cells.

Cytopathic effect   A pathological change that occurs in cells as a result of viral replication.

Cytoplasm   The area of the cell enclosed by the cell membrane containing organelles that function in cell metabolism and multiplication.

Cytotoxic T cells   Members of the T-cell subset designated CD8.

Cytotoxins   Exotoxins that damage or kill host cells.

$D = nV/R$ An equation representing the struggle between disease-producing microbes and host resistance. D is the severity of infection; n is the number of organisms; V is for virulence factors; R is for resistance factors.

DTaP vaccine   A newer DPT vaccine that utilizes acellular pertussis.

Debridement   Removal of necrotic (dead) tissue in an attempt to halt an infection.

Decomposers   Microbes that break down compounds into simpler constituents.

Defensive strategies   Adaptations that allow microbes to escape destruction by the host immune system.

Deletion   Removal of one (or more) nucleotide leading to a frame-shift change.

Dengue fever   A mosquito-borne disease that is typically self-limiting, with recovery occurring within about 10 days.

Dengue hemorrhagic fever   A potentially fatal disease that can result from infection with a dengue virus strain different from the one causing the initial infection.

Denitrifying   Returning nitrogen to the atmosphere.

Deoxyribonucleic acid   *See* DNA.

Dermotropic viruses   Viruses that have an affinity for and cause diseases in the skin and subcutaneous tissues.

Diatom   A type of unicellular alga.

Differential count   A reflection of the ratio of the white blood cell categories.

Diffusion   The passive movement of a substance from an area of high concentration to an area of low concentration.

DiGeorge syndrome   An immune disorder resulting from abnormal development of the thymus gland.

Dinoflagellate   A type of unicellular alga classified as a microbe; dinoflagellates are the primary source of food in the oceans.

Dioecious   Having distinct male and female forms in a species.

Diphtheria   An upper respiratory tract infection caused by *Corynebacterium diphtheriae*.

Diplococci   Groups of cocci that occur in pairs.

Direct transmission   Person-to-person contact in which the infectious agent is directly transferred from a portal of exit to a portal of entry.

Disease   A possible outcome of infection in which health is impaired in some fashion.

Distilled spirits   Alcoholic beverages resulting from bacterial fermentation and having a high alcoholic content, e.g., brandy, rum, and whiskey.

DNA   Molecules that store the genetic information.

Dose   The number of microorganisms to which a host has been exposed.

DOTS   Direct observational therapy short course, a strategy to ensure that individuals infected with tuberculosis take their prescribed medicines.

DPT vaccine   A triple vaccine against diphtheria, pertussis, and tetanus.

Dracunculiasis   A disease caused by the parasitic guinea worm, *Dracunculus medinensis*.

Duodenum   The first segment of the small intestine.

Eastern equine encephalitis virus   A common arbovirus found in the United States.

Ebola hemorrhagic fever   A severe viral infection characterized by extreme hemorrhaging with a high fatality rate.

Ecosystem   A population of organisms in a particular physical and chemical environment.

Ehrlichiosis   A bacterial tick-borne infection similar to Lyme disease.

Elephantiasis   A disease caused by filarial worms that results in blocked lymphatic vessels and the accumulation of large amounts of lymph fluid in the tissues.

Embryonated (fertile) chicken eggs   Chicken eggs containing live embryos; they are used to grow viruses.

Encephalopathy   A condition involving brain pathology as in transmissible spongiform encephalopathy.

Encystation   A process that allows protozoans to survive outside a host; the parasite is surrounded by a thick capsule.

Endemic disease   A disease that is continually present at a steady level in a population and poses little public threat.

Endemic relapsing fever   A tick-borne bacterial disease caused by *Borrelia recurrentis*.

Endemic typhus   A disease caused by *Rickettsia typhi*; it is transmitted to humans from the bite of rat fleas.

Endocytosis   A process of engulfment of material, including viruses, displayed by some cells.

Endotoxin   A toxin produced by some gram-negative bacteria; it is usually released not during cell growth but upon death and disintegration of the microbe.

Enology   The science of wine making.

Entero-Test   A method to test for the presence of *Giardia* trophozoites; the patient swallows a gelatin capsule attached to a string, and after 4 hours, the capsule is withdrawn and examined for the presence of trophozoites.

Enterotoxigenic *Escherichia coli* strains   Strains of *E. coli* that are a common cause of traveler's diarrhea.

Enterotoxin   A toxin that affects the intestinal tract.

Envelope   A bacterial structure consisting of a capsule, cell wall, and cell membrane.

Enveloped viruses   A category of viruses characterized by a membrane surrounding the capsid.

Enzymes   Substances that act as catalysts on specific substrates and influence the rate of a chemical reaction; some bacterial enzymes are virulence factors.

Eosinophils   White blood cells associated with helminth infections and allergies.

Epidemic   A disease that has a sudden increase in morbidity and mortality in a particular population.

Epidemic relapsing fever   A spirochete-caused disease transmitted from person to person by the bite of infected body lice.

Epidemic typhus   A disease caused by *Rickettsia typhi*; it is transmitted directly from human to human by the bites of body lice.

Epidemiology   The study of the sources, causes, and distribution of diseases and disorders that produce illness and death in humans.

Epstein-Barr virus   A DNA virus that is the primary cause of infectious mononucleosis.

Erythrocytes   Red blood cells.

Erythrogenic toxin   A secretion, produced by some strains of streptococci, that causes the red rash of scarlet fever.

Espundia   *See* Mucocutaneous leishmaniasis.

ETEC   *See* Enterotoxigenic *E. coli* strains.

Eucarya   One of three domains in the Woese system of classification.

Eucaryotic cells   Cells possessing a nuclear membrane and other membrane-bound organelles.

Excystation   The process by which a cyst comes out of dormancy resulting in an active stage.

Exfoliative toxin   A substance produced by staphylococci that causes the skin to become blistery and peel away.

Exotoxins   Protein molecules that are released by microbes during their growth and metabolism.

Extracellular microbes   Microbes that colonize on cell surfaces.

Extremophiles   Microbes that grow under harsh environmental conditions.

Extrusion   A process whereby mature virus particles are released from the host cell; also called budding.

Facultative anaerobes   Organisms that grow best in the presence of oxygen but are capable of survival in its absence.

Fermentation   A metabolic reaction regulated by enzymes that break down sugars to acids and carbon dioxide.

Filarial worms   Tiny adult threadlike worms that can block the lymphatic vessels, possibly resulting in elephantiasis.

Filariasis   *See* Elephantiasis.

Fixation   The process through which atmospheric nitrogen is converted to a product (typically ammonia) that plants are capable of using in metabolic processes.

Flagella   Structures composed of the protein flagellin that provide motility to certain species of bacilli and cocci.

Flagellin   The protein of which flagella are composed.

Flatworms   A morphological category of parasitic worms.

Flesh-eating strep   A streptococcus causing a severe form of streptococcal infection, necrotizing fasciitis.

Fomites   Inanimate objects that serve as means of transmission of infectious material.

Food infection   The result of ingestion of bacteria in contaminated foods and their subsequent growth in the intestinal tract accompanied by their secretion of toxin.

Food intoxication   The result of ingestion of bacterial toxins.

Foreign   Not normally present in the body and capable of triggering an immune response.

Fungi   A category of eucaryotic organisms including mushrooms and yeasts.

Fusion   A mechanism of penetration employed by some viruses in which there is contact between the viral envelope and the host cell membrane.

Gametes   Male or female sex cells.

Gamma globulin   A fraction of the globulin component of blood plasma in which most of the antibodies are present.

Ganglia   Collections of nerve cells.

Gas gangrene   A condition brought about through contamination of a wound with particles of soil containing *Clostridium perfringens* or its spores.

Gastrointestinal anthrax   A form of anthrax resulting from the ingestion of inadequately cooked meat contaminated with *Bacillus anthracis*.

Generation time   The length of time between rounds of binary fission.

Genes   Segments of the DNA molecule, involved in heredity, which encode specific polypeptide, protein, or RNA molecules.

Genetic engineering   A means of manipulating genes to bring about a desired function.

Genital warts   One of the most common sexually transmitted diseases, caused by human papillomaviruses.

Genome   The complete set of chromosomes in a cell.

Genus   A category in the binomial system of nomenclature above the species level.

Giardiasis   An intestinal disease caused by the protozoan *Giardia lamblia* and acquired by drinking contaminated water.

Glomerulonephritis   A kidney disease that is a potential result of streptococcal infection.

Gonorrhea   A sexually transmitted disease caused by *Neisseria gonorrhoeae*.

gp120   A molecule found on the surface of HIV that "docks" with CD4 receptor molecules on T lymphocytes.

Gram negative   A cell wall–staining property of some bacteria; they do not retain the crystal violet stain after decolorizing with alcohol and appear pink to red after staining with safranin.

Gram positive   A cell wall–staining property of some bacteria that causes them to retain the crystal violet stain after decolorizing with alcohol; they appear purple.

Granuloma   A pocket-like arrangement of cells associated with chronic inflammation that walls off the inflammatory agent.

Growth   An increase in the size of individual cells or an increase in a population of cells.

Guillain-Barré syndrome   A potentially fatal and rare nervous system disease of unknown cause; it sometimes occurs after administration of a vaccine.

Gummas   Tumorlike lesions that characterize tertiary syphilis.

H spike   *See* Hemagglutinin.

HAART   Highly active antiretroviral therapy; a cocktail of medications used to treat AIDS.

Halophiles   Bacteria that live in environments containing extreme salt concentrations.

Hanging drop   A method of observing bacterial motility under a microscope by suspending a drop of culture on a slide.

Hansen's disease   A disease caused by *Mycobacterium leprae*; formerly known as leprosy.

Hantavirus   A virus that causes hantavirus pulmonary syndrome.

Hantavirus pulmonary syndrome   A disease caused by a hantavirus that produces severe influenzalike respiratory problems and can result in death.

HBV   *See* Hepatitis B virus.

HCV   *See* Hepatitis C virus.

Healthy carriers   Individuals who have no symptoms of a particular microbial disease but harbor the microbes and may unwittingly pass the disease on to others.

Helical   Arranged in a continuous tube containing a virus's nucleic acid. Refers to arrangement of capsomeres.

Hemagglutinin   An enzyme on the spikes of influenza viruses, aiding the penetration of cells by the virus.

Hemolysins   Secretions produced by some microbes that destroy red blood cells through the destruction of cell membranes.

Hemolytic uremic syndrome   Kidney damage occurring primarily in young children as a result of bacterial toxins.

Hepatitis   Inflammation of the liver.

Hepatitis A virus   A virus found in feces and transmitted through contaminated drinking water and food. It is the most common hepatitis virus and usually causes only mild disease.

Hepatitis B virus   A virus transmitted in body fluids. It causes subclinical to severe disease.

Hepatitis C virus   A virus transmitted mainly by intravenous drug use; a major reason for liver transplants in the United States.

Hepatitis D virus   An incomplete virus that requires hepatitis B virus to replicate. It causes severe disease with a high mortality rate.

Hepatitis E virus   A virus transmitted by the fecal-oral route. It causes a disease that is usually of moderate severity.

Herbivorous   Refers to organisms that feed on plants.

Herd immunity   Immunity of enough members of a population to protect against an epidemic.

Hermaphroditic   Producing both sperm and eggs.

Herpes simplex virus   A virus that causes the highly infectious disease herpes. Herpes simplex virus type 1 causes painful sores around the mouth and lips, frequently referred to as cold sores or fever blisters. Herpes simplex virus type 2 causes painful sores on the penis in males and on the labia, vagina, or cervix in females, typically referred to as genital herpes.

Heterotrophs   Organisms that require organic compounds as an energy source.

HGE *See* Human anaplasmosis.

Hiker's diarrhea *See* Giardiasis.

HIV *See* Human immunodeficiency virus.

HIV seropositive Having antibodies against human immunodeficiency virus.

HME *See* Human monocytic ehrlichiosis.

Homologous Refers to streches of DNA with identical or closely related sequences.

Hookworm disease A disease that is caused by the roundworms *Ancylostoma duodenale* and *Necator americanus*.

Horizontal gene transfer A process by which genes are transferred from one organism to another.

Horizontal transmission A means by which a disease spreads from one person to another.

HPV *See* Human papillomaviruses.

HSV *See* Herpes simplex virus.

Human anaplasmosis A tick-borne bacterial infection caused by an unknown species of *Ehrlichia*.

Human Genome Project The mapping of the genes located on the 23 pairs of human chromosomes.

Human growth hormone A hormone that is used for the treatment of dwarfism and is now produced by genetic engineering.

Human immunodeficiency virus The virus that causes AIDS. Human immunodeficiency virus type 1 is the most common cause of AIDS worldwide, and human immunodeficiency virus type 2 is the most common cause of AIDS in West Africa.

Human insulin A hormone that regulates the amount of glucose in the bloodstream. Much of the insulin now used to treat diabetes is a product of genetic engineering.

Human monocytic ehrlichiosis A tick-borne bacterial infection caused by *Ehrlichia chaffeensis*.

Human papillomaviruses Viruses that cause warts.

Hyaluronidase An enzyme produced by some bacteria that breaks down hyaluronic acid found in connective tissue.

Hydrophobia Fear of water; an old term for rabies.

Hyperthermophiles Bacteria that live in extremely hot environments.

Hyphae Intertwined filaments characteristic of molds.

Iatrogenic infection Infection induced in a patient by a medical procedure.

Icosahedrons Three-dimensional, 20-sided structures with triangular sides; they confer a geodesic shape to a virus.

ID *See* Infective dose.

Ig *See* Immunoglobulins.

Illness stage The time in which disease develops to the most severe stage, as evidenced by typical signs and symptoms.

Immunocompromised Characterized by a weakened immune system as a result of AIDS or other conditions.

Immunoglobulins Categories of antibodies (immunoglobulins A, D, E, G, and M), each possessing specific properties.

Impetigo A superficial infection of the skin that is typically manifested by blisters around the mouth and is usually caused by staphylococci.

Incubation stage The time between a pathogen's access into the body and the appearance of signs and symptoms.

Indirect transmission The passage of infectious material from a reservoir to an intermediate host and then to a final host.

Inducible enzymes An enzyme that is the result of genes that can be turned on or off depending upon the circumstances.

Infantile paralysis A form of poliomyelitis in infants and young children, characterized by muscle paralysis.

Infection The presence of microbes in the body without definitive symptoms.

Infectious agents Plasmids capable of exchanging genetic material from one microbe to another; some confer antibiotic resistance.

Infective dose The minimum amount of bacteria needed to establish an infection.

Infectious mononucleosis A disease caused by Epstein-Barr virus in which the salivary glands are infected by the virus; frequently referred to as "mono."

Inflammation Swelling, pain, redness, and heat in an area of tissue.

Influenza Exaggerated coldlike symptoms caused by influenza virus.

Influenza A virus A strain of influenza virus that causes epidemics and occasionally pandemics.

Influenza B virus A strain of influenza virus that causes epidemics and does not have an animal reservoir.

Influenza C virus A strain of influenza virus that does not produce epidemics and causes mild respiratory illness.

Inhalation anthrax The most severe form of anthrax, resulting from the intake of anthrax spores.

Inorganic compound Generally, a chemical compound that does not contain carbon (although carbon dioxide and several other carbon-containing compounds are considered inorganic).

Insects A large class of arthropods characterized by three body segments and six legs.

Insertion Addition of one (or more) nucleotides leading to a frame-shift change.

Interferon A component of blood that interferes with viral replication.

Internally displaced persons Persons or groups of persons for whom social disruptions have forced to move but have not crossed international borders.

Internationally displaced persons Persons or groups of persons for whom social disruptions have forced to move across international borders.

Intracellular bacteria   Bacteria that penetrate and grow inside host cells.

Iron lung   A barrel-like device formerly used for persons who had difficulty breathing on their own because of muscle paralysis as a result of polio.

Ivermectin   A drug used to treat onchocerciasis.

Kala-azar   A severe and usually fatal form of leishmaniasis in which the parasites invade the liver and other organs.

Kaposi's sarcoma   A type of vascular cancer that occurs in people suffering from AIDS.

Kinases   Enzymes that break down clots in the blood.

Kissing bug   A triatomid insect that is the vector for Chagas disease.

Koch's postulates   A set of rules used to establish that a particular organism is the cause of a particular disease.

Koplik's spots   Spots that appear in the mouth in the early stage of measles.

Kuru   A transmissible spongiform disease formerly found among members of the Fore people of New Guinea.

Larva   A stage in arthropod development that occurs after hatching and before maturity is reached.

Laryngitis   Inflammation of the larynx.

Latency   The period in which a microbe is in the host without displaying any visible symptoms.

LD50   50% lethal dose; a laboratory measurement of virulence to determine the dose that kills 50% of the test animals in a given time.

Legionnaires' disease   An airborne pneumonialike disease caused by *Legionella pneumophila*.

Leguminous plant   A type of plant with swellings or nodules along the root system containing *Rhizobium* and other nitrogen-fixing bacteria.

Leishmaniasis   A disease caused by parasites and transmitted by the bite of female sand flies.

Lepromas   Tumorlike skin lesions seen in leprosy.

Leprosy   *See* Hansen's disease.

Leptospirosis   A disease caused by the spirochete *Leptospira interrogans*.

Leukemia   A type of cancer characterized by uncontrolled reproduction of white blood cells.

Leukocidins   Bacterial enzymes that destroy white blood cells.

Leukocytes   White blood cells.

Lipopolysaccharide   A molecule with both a lipid and a polysaccharide component found in the outer wall of gram-negative bacteria; it acts as an endotoxin.

Lockjaw   A symptom of tetanus characterized by contraction of the muscles in the jaw; also an alternative name for tetanus.

Lower respiratory tract infections   Infections that occur in the trachea, larynx, bronchi, bronchioles, or lungs.

Lymph   A tissue fluid derived from blood that is returned to the blood by lymphatic vessels.

Lymph nodes   Structures along lymphatic vessels that act as microbe filters.

Lymphadenopathy   Swelling of lymph nodes that occurs in AIDS and other infections.

Lymphatic system   A system in which lymph is transported in lymphatic vessels.

Lymphatic vessels   Structures that transport lymph fluid.

Lymphocytes   A category of white blood cells that play a key role in immunity.

Lymphocytic leukemia   A form of leukemia characterized by overproduction of lymphocytes, leading to abnormally large numbers of immature and nonfunctional lymphocytes.

Lymphoid path   The development of blood cells that leads to the production of lymphocytes.

Lysis   Bursting of cells.

Lysogenized   Refers to bacterial cells that contain phage nucleic acid.

Lysozyme   An enzyme present in tears that contributes to the nonspecific immune system by disrupting bacterial cell walls.

M protein   A protein found in the cell walls of streptococci that confers resistance to phagocytosis.

Macrophages   Phagocytic cells that function in the immune system.

Macroscopic   Visible without the aid of a microscope.

Mad cow disease   A disease of cattle caused by prions that results in spongy degeneration of the brain accompanied by severe and fatal neurological damage.

Major histocompatibility molecules   Molecules associated with presentation of antigenic material to receptor molecules on cytotoxic T cells.

Malaria   A tropical disease caused by the bite of female *Anopheles* mosquitoes infected with the *Plasmodium* protozoan parasite.

Mastigophora   One of the four groups of protozoans, characterized by the presence of flagella.

MDR tuberculosis   Multi-drug resistant strains of tuberculosis.

Mechanical vectors   Arthropod vectors that transmit microbes passively on their body parts; the microbes do not invade, multiply, or develop in the vector.

Memory cells   B cells that do not progress to antibody-producing plasma cells but are retained for a quick immune response in the event of future exposure to the same antigen.

Meninges   Membranes around the spinal cord and the brain.

Meningitis   Inflammation of the meninges that can be caused by microbes.

Merozoites   Forms in the malaria life cycle resulting from the asexual multiplication of sporozoites.

Metastasis   The spreading of cancer cells from one region or tissue to another.

Miasma   A term, used before microbes were identified as agents of disease, meaning "bad air" or "swamp air."

Microfilariae   Mature filarial worms in the blood that can be taken up by mosquitoes to infect other people with elephantiasis.

Micrometer   A unit of measurement equal to one millionth of a meter.

Microscopic   Visible only with the aid of a microscope.

Missense mutation   A change in a codon resulting in incorporation of a different amino acid.

MMR vaccine   Measles, mumps, and rubella vaccine.

Monocytes   A category of phagocytic white blood cells.

Mononucleosis   *See* Infectious mononucleosis.

Morbidity   A measure of the rate of illness of a particular disease.

Mortality   A measure of the rate of death resulting from a particular disease.

MRSA   Methicillin-resistant strains of *Staphylococcus aureus.*

Mucocutaneous leishmaniasis   A form of leishmaniasis characterized by invasion of the parasite into the skin and mucous membranes, causing destruction of the nose, mouth, and pharynx; also called espundia.

Multicellular   Having more than one cell.

Multiple drug resistant   Resistant to more than one antibiotic.

Multiplication   A process resulting in an increase in the total number of cells.

Murine typhus   A variety of typhus fever caused by *Rickettsia typhi* that is transmitted by fleas.

Mutagens   Physical or chemical agents that cause mutation

Mutation   A change in DNA that is transferred to subsequent generations.

Mutualism   A symbiotic relationship in which both organisms benefit from the association.

Mycoplasmas   Bacteria that have no cell walls.

Myeloid leukemia   A type of cancer resulting from overproduction of monocytes and granular leukocytes.

Myeloid path   The lineage of blood cell maturation that leads to the production of platelets, red blood cells, monocytes, neutrophils, eosinophils, and basophils.

N spike   *See* Neuraminidase.

Naked viruses   Viruses that are not enclosed by an envelope.

Nanometer   A unit of measurement equal to one billionth of a meter.

Narrow spectrum antibiotic   An antibiotic that is inhibitory to limited range of bacteria.

Necrotic   Dead; used in reference to tissue.

Necrotizing fasciitis   A condition caused by highly invasive streptococci in which the subcutaneous tissue is infected; the streptococci are sometimes referred to as "flesh-eating."

Negri bodies   Cytopathic structures found in rabies virus-infected cells.

Neonatal (newborn) tetanus   A manifestation of tetanus in newborn children that results from unsanitary conditions during delivery.

Neuraminidase   An enzyme on the spikes of some influenza viruses, aiding in the release of new virions.

Neurocysticercosis   The presence of cysticerci in the brain or spinal cord resulting in seizures, headaches, and possibly death.

Neurosyphilis   A late stage of syphilis involving neurological damage possibly characterized by paralysis and insanity.

Neurotoxins   Toxins released by microbes that interfere with the transmission of neural impulses.

Neutrophils   White blood cells that are phagocytic.

Nitrification   A stage in the nitrogen cycle in which ammonia is converted into nitrogen.

Nitrogen cycle   A cycle in nature in which atmospheric nitrogen is recycled through a pathway involving bacteria.

Nonenveloped viruses   Viruses that do not have an envelope around them; also referred to as "naked."

Nonsense mutation   A change in a codon leading to a stop codon.

Nonspecific immunity   Physiological defenses that prevent microbes from gaining access into the body or eliminate those that have penetrated the body.

Norwalk virus   A virus that is a frequent cause of gastroenteritis in older children and adults.

Norwalk-like viruses   A group of viruses, related to Norwalk virus, that frequently cause gastroenteritis in older children and adults.

Nosocomial infections   Infections acquired by patients during hospitalization or during stays in long-term health care facilities.

Nuclein   An early term for DNA.

Nucleocapsid   A viral structure consisting of the viral nucleic acid and the protein coat.

Nucleoid   The DNA-rich area in procaryotic cells; it is not surrounded by a membrane.

Nutrient agar   A semisolid growth medium used to grow bacteria in the laboratory.

Nymph   A preadult stage in tick development.

Obligate intracellular parasites   Microbes that can only replicate within cells.

Offal   A ground-up mixture of organs and trimmings of dead animals used to feed other animals.

Offensive strategies   Adaptations by microbes that result in their ability to damage the host and establish disease.

Onchocerciasis   A parasitic disease caused by *Onchocerca volvulus* that frequently results in blindness.

Oocyst   Infectious form of *Cryptosporidium parvum.*

Operons A group of functionally related genes that act as "on and off" switches.

Ophthalmia neonatorum A condition in which the corneas are damaged as a result of the transmission of *Neisseria* into the eyes of newborns during delivery.

Opisthotonos A body position in which the back is severely arched; characteristic of tetanus.

Opportunistic pathogens Organisms that cause disease when the host's immune system is weakened, as in AIDS.

Oral rehydration therapy A method of cholera treatment designed to replace lost body fluids; an alternative to intravenous rehydration.

Orchitis Inflammation of the testes that sometimes occurs in males infected with mumps virus.

Organ system A collection of organs that contribute to an overall function.

Organic compound Generally, a chemical compound that contains carbon (although carbon dioxide and several other carbon-containing compounds are considered inorganic).

Organisms Living entities composed of cells.

Organs Structures composed of more than one tissue type.

Oriental sore A form of leishmaniasis that results in skin lesions primarily on exposed parts of the body; also known as cutaneous leishmaniasis.

Oseltamivir A relatively new drug that shortens the duration of influenza caused by influenza A and B viruses. The trade name of this drug is Tamiflu.

Osmosis The movement of a solvent from an area of low solute (high solvent) concentration to an area of high solute (low solvent) concentration.

Outer membrane An additional layer external to the peptidoglycan layer in Gram-negative bacteria that is associated with virulence.

Pandemic A worldwide outbreak of disease.

Paralytic poliomyelitis A form of polio caused by replication of poliovirus in nerve cells, sometimes resulting in severely deformed limbs and paralysis.

Parasitic cycle A chain of events, sometimes quite complex, by which a parasite exits from a host and gains access to a new host.

Parasitism A form of symbiosis in which one biological agent lives at the expense of another.

Parotid glands One of the three pairs of salivary glands; these glands may become infected and cause swelling on one or both sides of the face, a characteristic of mumps.

Parthenogenesis A process by which females produce eggs without fertilization by males.

Pathogens Microbes capable of producing disease.

Pellicle A protective, rigid cover outside the cell membrane found in some protozoans.

Pelvic inflammatory disease A condition that occurs in about 50% of untreated female gonorrhea patients characterized by abdominal pain and sometimes sterility.

Penetration A stage in the replication of viruses in which the virus enters the host cell.

Peptidoglycan A compound found in Gram-positive and Gram-negative bacterial cell walls that confers rigidity and tensile strength.

Perforin A membrane-penetrating protein released by cytotoxic T cells that leads to lysis of cells harboring viruses.

Peritonitis A potentially fatal condition caused by leakage of intestinal fluids into the abdomen.

Petri dish A round, dishlike container in which bacteria are grown on agar in the laboratory.

Phage conversion A process by which bacteriophage DNA is incorporated into the bacterial chromosome and confers new properties on the bacterial host.

Phagocytic cells Cells that engulf microbes and bring about their destruction by enzymatic activity; these cells play a vital role in the nonspecific immune system.

Phagocytosis A nonspecific body defense mechanism by which bacteria are ingested and killed by cells.

Photosynthetic autotrophs Organisms that utilize the energy of the sun and that use carbon dioxide as a carbon source.

PID *See* Pelvic inflammatory disease.

Pili Bacterial appendages in gram-negative bacteria that act as adhesins; some serve as a bridge allowing genetic exchange.

Pilin The protein that composes pili.

Pinworm A common helminthic disease caused by the worm *Enterobius vermicularis*.

Plague A bacterial disease caused by *Yersinia pestis*, a highly virulent bacterium; also known as the Black Death.

Plankton The primary food source for many aquatic organisms, consisting mainly of protozoans.

Plantar warts Deep, painful warts found on the soles of the feet; caused by human papillomavirus.

Plasma cells Antibody-producing cells derived from B cells.

Plasmids Small molecules of nonchromosomal DNA found in some bacteria.

Platelets Blood cells that initiate the clotting of blood.

Pleurisy Inflammation of the pleural lining of the lungs.

Plug drugs Drugs that act to both prevent and treat influenza by interfering with the N spikes of the virus.

Pneumonia Inflammation of the lungs; can be caused by a variety of microbes.

Pneumonic plague A form of bubonic plague that develops into pneumonia.

Pneumotropic viruses Viruses that have an affinity for the respiratory tract.

Point mutation A genetic process in which one nucleotide is replaced with another nucleotide; simplest mutation.

Polio Poliomyelitis, a highly infectious viral disease that can lead to muscle paralysis.

Polyhedral Consisting of icosahedra; a term used to describe an arrangement of viral capsomeres.

Polypeptide A series of amino acids chemically held together by peptide bonds.

Pontiac fever A pneumonialike illness caused by *Legionella pneumophila.*

Portal of entry The site at which microbes enter a host.

Portal of exit The site from which microbes leave a host and may infect another host.

PPD *See* Purified protein derivative.

Primary immune structures The bone marrow and the thymus gland in which maturation of B and T lymphocytes takes place.

Primary producers Photosynthetic organisms that utilize the energy of the sun and produce organic compounds and oxygen.

Primary syphilis An early stage of syphilis characterized by the formation of chancres on the genitals.

Prions Infectious, highly stable, misfolded proteins that can cause neurological disease; they do not contain DNA or RNA.

Procaryotic Not having the genetic material contained within a membrane in the cell.

Prodromal stage An early stage in a microbial disease characterized by headache, tiredness, and muscle aches. Also the first stage of human immunodeficiency virus infection characterized by fever, diarrhea, rash, aches, fatigue, and lymphadenopathy.

Proglottids Compartmentlike segments of tapeworms containing both testes and ovaries.

Promoter site A site on a gene strand that marks the beginning of transcription.

Propagated epidemics Epidemics resulting from direct person-to-person transmission.

Prophage A segment of phage integrated into a bacterial chromosome.

Protease inhibitors A class of drugs used in AIDS therapy that prevent viral replication.

Protozoans A category of unicellular eucaryotic microbes.

PrP gene Gene located on chromosome 20 that is responsible for prion proteins.

Psychrophiles Cold-loving bacteria that grow best at about 15° Celsius.

Purified protein derivative A product derived from *Mycobacterium tuberculosis* and used in the tuberculin skin test.

R (resistance) factors Genes carried on plasmids that confer antibiotic resistance.

Rabies A viral disease transmitted by the bite of a rabid animal, resulting in damage to the nervous system and eventual death if not treated promptly by vaccination; formerly called hydrophobia.

Recombinant DNA technology *See* Genetic engineering.

Red blood cells Cells of the circulatory system that carry oxygen throughout the body; also called erythrocytes.

Release The last stage in the viral replication cycle during which mature viruses are released from host cells.

Relenza *See* Zanamivir.

Rendering process The process by which animal parts are turned into animal feed.

Replication The process by which viruses multiply in host cells.

Reservoir A site in nature where microbes survive and multiply and from which they may be transmitted.

Resistance A defensive function of the immune system affording protection against microbial invasion; also, the ability of microbes to counter the effects of antimicrobial agents.

Respiratory syncytial virus A prevalent cause of respiratory illness, most commonly found in infants, that produces non-specific symptoms, including fever, runny nose, ear infection, and pharyngitis.

Rheumatic fever A condition involving the heart and joints resulting from repeated bouts of streptococcal infection.

Rhinoviruses A large group of viruses that are responsible for many common colds.

Ribavirin A drug used in the treatment of hepatitis C and in the treatment of pneumonialike illnesses caused by respiratory syncytial virus.

Ribosomal ribonucleic acid A type of ribonucleic acid (RNA) that is associated with ribosomes.

Rickettsiae Rod-shaped bacteria that are (with a single exception) transmitted through the bite of arthropods; they are intracellular obligate parasites.

River blindness A common type of blindness resulting from the migration of larval forms of the worm *Onchocerca volvulus* into the eye; also called onchocerciasis.

Rocky Mountain spotted fever A tick-borne disease common in the southeastern United States that is caused by *Rickettsia rickettsii.*

Ropy milk A condition that occurs when *Alcaligenes viscolactis* sheds its slime layer into milk.

RotaShield An oral vaccine against rotavirus; the vaccine was withdrawn about a year after its approval because of adverse reactions.

Rotaviruses A group of viruses that are a common cause of viral gastroenteritis in children under the age of 5 years.

Roundworms A morphological category of worms.

Rubella A viral disease characterized by a rash on the face that spreads to the trunk and the extremities.

Saber shins A condition sometimes seen in syphilis patients in which the shinbone develops abnormally.

Sabin vaccine A polio vaccine developed by Albert Sabin consisting of attenuated polioviruses.

Salk vaccine The first vaccine against polio developed by Jonas Salk containing inactivated poliovirus.

Salmonellosis   A condition caused by ingestion of salmonella bacteria resulting in gastroenteritis manifested by nausea, vomiting, abdominal cramps, and diarrhea.

Sand flies   Insects that transmit *Leishmania* species.

Sarcodina   A group of protozoans that exhibit a "creeping" movement.

Scalded skin syndrome   A condition caused by a staphylococcal toxin that results in the skin's becoming blistery with a tendency to peel.

Scarlet fever   A disease caused by a strain of streptococcus that produces an erythrogenic toxin leading to the development of a red rash and a strawberry-colored tongue.

Scavengers   A synonym for decomposers.

Schistosomiasis   A parasitic disease caused by blood flukes that enter the body through the skin.

SCID   *See* Severe combined immunodeficiency.

Scolex   The head of a tapeworm; it attaches to the host intestinal wall by suckerlike projections.

Secondary syphilis   A stage of syphilis characterized by a rash on the palms and soles; during this stage, the spirochetes multiply and spread throughout the body.

Secondary bacterial infections   Infections that result from bacteria in individuals suffering from the flu or other conditions that lower immune resistance.

Secondary immune structures   The spleen, tonsils, adenoids, lymph nodes, and patches of tissue associated with the intestinal tract that are seeded with mature B and T cells and phagocytic cells.

Selectively permeable   Able to be permeated by some but not all types of molecules; a property of cell membranes.

Septicemic plague   A variety of plague resulting from the spread of infection from the lungs to other parts of the body.

Serum sickness   A condition characterized by the formation of antigen-antibody complexes that are deposited in the skin, kidney, and other sites, as might occur in treatment with immunoglobulin.

Severe combined immunodeficiency   A disease in which individuals lack both functional T and B cells and cannot mount either an antibody- or a cell-mediated immune response.

Sex pili   Structures that function in exchange of DNA between bacteria by forming a bridge between cells.

Sexual stage   A stage of malaria in which sporozoites are produced from *Plasmodium* parasites in the blood.

Sexually transmitted diseases   Diseases caused by transmission of microbes from the warm, moist mucous membranes of one individual to the mucous membranes of another individual during sexual contact.

Shigellosis   A bacterial gastrointestinal illness caused by the ingestion of *Shigella*-contaminated foods and water resulting in symptoms of diarrhea, abdominal cramping, and, in some cases, dysentery.

Shingles   A disease caused by varicella-zoster virus that occurs in some individuals with a history of chicken pox; the virus infects nerve fibers, creating intense pain.

Sleeping sickness   A protozoan disease caused by the bite of a tsetse fly carrying the parasite. Two types (West and East African trypanosomiasis) are known.

Slime layer   A heavy mucuslike material that accumulates around some bacteria.

Small animalcules   The name for microbes first described by Antony van Leeuwenhoek during his examination of tooth scrapings with a primitive microscope.

Species   The fundamental rank in the binomial system of classification of organisms; in the name *Escherichia coli*, "*coli*" refers to the species.

Specific immunity   Immunity that is acquired after birth and responds to the presence of foreign and potentially harmful microbes that have breached the external and nonspecific defense mechanisms.

Spikes   Projections that extrude through the surface of some viruses.

Spirilla   Spiral-shaped bacteria.

Spirochetes   Flexible, corkscrew-shaped, motile bacteria.

Spleen   An organ that is part of the secondary immune system; it contains phagocytic cells and both mature B and T cells.

Spontaneous generation   A false but once popular theory that nonliving structures could give rise to living organisms. Louis Pasteur played a major role in disproving this theory.

Sporadic   Occurring occasionally and at irregular intervals in a random and unpredictable fashion.

Spores   Structures contained within dormant bacterial cells that are highly resistant to heat, drying, radiation, and a variety of chemical compounds. The genera *Bacillus* and *Clostridium* are sporeformers.

Sporozoites   Forms of *Cryptosporidium parvum* that penetrate the intestinal cells; also, infective forms of malaria parasites.

St. Louis encephalitis virus   A common mosquito-borne arbovirus found in the United States that causes encephalitis.

Stage of decline   The time following an illness in which symptoms begin to disappear and the body returns to normal.

Staph food poisoning   An illness caused by the secretion of an enterotoxin from *Staphylococcus aureus;* characteristic symptoms are abdominal cramps, nausea, vomiting, and diarrhea.

Staphylococci   Bacteria of the genus *Staphylococcus* that group together in clusters resembling bunches of grapes.

Stem cells   Undifferentiated cells capable of differentiating into specialized cells.

Stomach flu   A common but meaningless term that is associated with gastroenteritis.

Strep throat   A mild airborne infection caused by *Streptococcus pyogenes* with characteristic symptoms of red or sore throat, fever, and headache.

Streptococci   Bacteria of the genus *Streptococcus* that group together in chains resembling strings of pearls.

Streptokinase   A product of streptococci that dissolves blood clots.

Stromatolites   Fossilized mats formed from microorganisms and dating as far back as 3.5 billion to 3.8 billion years.

Strongyloidiasis   Illness caused by the nematode *Strongyloides stercoralis;* symptoms are nausea, vomiting, anemia, weight loss, and chronic bloody diarrhea.

Subclinical   Asymptomatic and not diagnosed.

Sucking disks   Structures by which *Giardia lamblia* trophozoites attach to the lining of the intestine.

Swamp fever   *See* Leptospirosis.

Swimmer's itch   *See* Cercarial dermatitis.

Symbiosis   A relationship between two or more organisms that live together; mutualism, commensalism, and parasitism are all forms of symbiosis.

Syphilis   A sexually transmitted disease caused by *Treponema pallidum;* the symptoms are sores on the penis or cervix, rash, and degeneration of organs and tissues.

Systemic infections   Blood-borne infections that spread throughout the body.

T-cell subset   A collective term for the categories of T cells differentiated by clusters of differentiation.

T-helper cells   Members of the T-cell subset designated CD4.

T lymphocytes   A subcategory of lymphocytes that function in the immune system in several ways.

Tamiflu   *See* Oseltamivir.

Taxonomy   The science of classification of organisms.

Terminator sequence   A site on DNA marking the end point of transcription.

Tertiary syphilis   The third stage of syphilis during which organs and tissues undergo degenerative changes.

Tetanospasmin   A neurotoxin produced by *Clostridium tetani* that results in rigid paralysis.

Tetanus   A bacterial disease acquired by exposure to *Clostridium tetani* or its spores. Symptoms include stiffness in the jaw and contraction of muscles in the limbs, stomach, and neck. Also called lockjaw.

Tetrads   Groupings of cocci in clusters of four.

Thymus gland   The organ in which T-cell maturation is completed; it is located behind the sternum and just above the heart.

Tissue   A group or collection of cells that are all of the same type.

Tonsils   Structures of the secondary immune system located at the back of the throat that aid in protection from microbes entering through the nose and throat.

Toxemia   An illness resulting from the presence of an exotoxin in the body.

Toxic shock syndrome   A condition caused by certain strains of toxin-producing staphylococci; primarily associated with the use of highly absorbent tampons.

Toxigenicity   The ability of microbes to produce toxins.

Toxins   Major virulence factors that are harmful to the body and are produced by many pathogenic microbes; tetanus, botulinum, and erythrogenic toxins are examples.

Toxoplasmosis   A protozoan disease caused by *Toxoplasma gondii* and acquired by the ingestion of oocysts present in cat feces. Symptoms include sore throat, low-grade fever, and lymph node enlargement.

Tracheitis   Inflammation of the trachea.

Tracheotomy   The cutting of a hole in the throat to facilitate breathing.

Transcription   A genetic process in which DNA is "read" into mRNA.

Transduction   A recombinational process characterized by bacteriophage-mediated transfer of DNA.

Transfer RNA (tRNA)   A clover leaf-shaped molecule that transfers a single specific amino acid to the ribosome to build a polypeptide bond.

Transformation   A recombinational process characterized by the uptake of "naked" DNA into competent cells.

Transforming principle   A term used by Frederick Griffith to describe the transfer of virulence from virulent to nonvirulent bacteria; the active component was later discovered to be DNA.

Translation   A genetic process in which the mRNA nucleotide sequence is converted into an amino acid sequence, which is assembled into a protein.

Transmissible spongiform encephalopathies   Degenerative brain diseases that are thought to be the result of abnormally folded prions that latch onto normal prions and convert them into an altered, defective form.

Transmission   A link in the cycle of microbial disease between reservoir and portal of entry.

Transovarial transmission   The passage of microbes from adult ticks to their eggs.

Transposons   Also called "jumping genes," are genetic elements that move from one site on a chromosome to another or from a chromosome to a plasmid or from a plasmid to a chromosome.

Traveler's diarrhea   An illness caused by enterotoxigenic *Escherichia coli* strains.

Treponemes   Spirochetes released by chancres in individuals suffering from syphilis.

Triatomid insect   The vector for Chagas' disease, commonly referred to as the kissing bug.

Trichinellosis   A roundworm disease caused by *Trichinella spiralis* that is transmitted in undercooked pork. Symptoms include nausea, diarrhea, vomiting, fatigue, fever, headaches, chills, aching joints, and itchy skin.

Trichomoniasis   A sexually transmitted disease caused by *Trichomonas vaginalis.* Symptoms include intense itching, urinary frequency, pain during urination, and vaginal discharge in females. In males, symptoms include pain during urination, inflammation of the urethra, and a thin, milky discharge.

Trophozoite   The reproductive and feeding stage of parasitic amoebae and other protozoan parasites.

Tuberculin skin test (Mantoux test)   A method for diagnosis of tuberculosis; purified protein derived from *Mycobacterium tuberculosis* is injected into the arm of a patient, and the presence of an induration between 48 and 72 hours after injection indicates past or present exposure to the tubercle bacillus.

Tuberculosis   A contagious lower respiratory tract disease caused by *Mycobacterium tuberculosis*. Symptoms include fever, night sweats, weight loss, fatigue, and coughing up of blood-tinged sputum.

Typhoid fever   A disease caused by *Salmonella typhi* that is transmitted by flies and fomites.

Typhoid Mary   A nickname for Mary Malone (1869–1938), a notorious carrier of typhoid fever.

Ulcer   A lesion (sore) on a soft part of the body including the skin, duodenum, or stomach.

Unicellular   Having only one cell.

Upper respiratory tract infections   Infections occurring in the tonsils or pharynx.

Urea   A waste product of protein metabolism present in urine and other body fluids.

Urease   An enzyme that digests urea.

Variant Creutzfeldt-Jakob disease   A fatal human transmissible spongiform encephalopathic disease caused by prions. It is similar to mad cow disease in that it causes degeneration of brain tissue and severe neurological damage.

Varicella-zoster virus   The DNA virus that causes shingles and chicken pox.

VD   *See* Venereal diseases.

Vector borne   Carried by a particular organism.

Vegetative cells   Spore-forming bacteria without spores.

Venereal diseases   The earlier name for sexually transmitted diseases, commonly referred to as VD.

Vertical transmission   A method of transmission characterized by passage of pathogens from parent to offspring across the placenta, in breast milk, or in the birth canal.

Vesicle   A membrane-bound structure in the cytoplasm; also, a fluid-filled lesion that appears on the skin.

Vibrios   Spirillar bacteria in the shape of comma-curved rods; *Vibrio cholerae*, the causative agent of cholera, is an example.

Virion   A complete viral particle.

Virulence   The capacity of microbes to produce disease as a result of defensive and offensive strategies.

Viruses   One of the categories of microbes; characterized as subcellular obligate intracellular parasites.

Visceral leishmaniasis   The most severe form of leishmaniasis; symptoms are fever, weakness, weight loss, anemia, and protrusion of the abdomen. Also known as kala-azar.

Viscerotropic viruses   Viruses that affect internal organs (viscera) such as the liver, spleen, and intestines.

Wandering macrophages   Macrophages derived from monocytes of the blood that are phagocytic and move freely about the tissues.

West Nile virus   An arbovirus transmitted by the bite of a mosquito that emerged in New York in 1999; it causes encephalitis.

Western equine encephalitis virus   A common arbovirus in the United States.

White blood cells   Cells that play a critical role in body defense mechanisms; also called leukocytes. Lymphocytes, neutrophils, monocytes, basophils, and eosinophils are the five types of white blood cells.

Whoop   The characteristic sound of whooping cough; deep and rapid inspirations in the partially obstructed passages are responsible for the sound.

Whooping cough   A highly infectious disease caused by *Bordetella pertussis*. Symptoms include spasms of violent hacking and persistent, recurrent coughing with a whooplike noise. Also called pertussis.

XDR tuberculosis   Extreme drug resistant strains of tuberculosis.

Xenodiagnosis   A diagnostic method used in Chagas disease. "Kissing bugs" free of trypanosomes are allowed to feed on individuals suspected of having the disease. A few weeks later, the insects are examined for the presence of the parasite.

Yellow fever   A mosquito-borne viral disease also known as yellow jack; symptoms include fever, bloody nose, headache, nausea, muscle pain, vomiting, and jaundice.

Zanamivir   An antiflu drug that is effective against both A and B influenza viruses. It reduces the severity of the flu if taken within 48 hours of the appearance of symptoms. The trade name of this drug is Relenza.

Zidovudine   *See* AZT.

Zoonoses   Diseases for which domestic and/or wild animals are the reservoirs and which can be transmitted to humans.

Index

Boxes, figures, and tables are indicated with b, f, *and* t *following the page number.*